Tom Bowser grew up on his parents' farm, Argaty, in Perthshire. After university he worked as a teacher and then as a journalist before returning to Argaty in 2009. In 2017 he took over the running of the conservation project Argaty Red Kites and in 2018 launched the Argaty Red Squirrels project.

Gordon Buchanan is one of Britain's most popular wildlife film makers and presenters. His documentaries include *The Bear Family and Me* and BBC TV's *Into the Wild*.

'A wonderful book'
Euan McIlwraith, BBC TV's *Landward*

'An inspirational real-life story'
Scots Magazine

'Imbued with the author's passion for nature, and his desire to dedicate his family's land to conservation'
The Bookseller

'An environmental story that gives hope, a story of a bird that was extinct doing really well on an estate that's trying to promote a different way of doing things'
Dundee Courier

T0258355

A SKY FULL OF KITES

A Rewilding Story

Tom Bowser

BIRLINN

To my wife and best friend, Sarah, and our daughter,
Rowan, who brings us both so much joy. May your
skies always be filled with kites.

This edition first published in 2024 by
Birlinn Limited
West Newington House
10 Newington Road
Edinburgh
EH9 1QS

www.birlinn.co.uk

First published in 2021

ISBN 978 1 78027 776 9

A catalogue record for this book is available from the British Library

Typeset by Initial Typesetting Services, Edinburgh

Printed and bound by Clays Ltd, Elcograf S.p.A.

Contents

What Next?

Author's Note to the 2024 Edition

Revisiting one's own book is a strange experience for any author. Like everyone else, we are forever changing as we age and grow. A book can, therefore, never be considered a statement of who you will always be, merely of who you were at the time of its writing. Returning to *A Sky Full of Kites* now, I see how much has changed in the few short years since it was published. I see how much I have changed.

This was a story of new beginnings, for me and for Argaty. In those years I was setting off on a journey, trying to understand who I was and what I was capable of. Argaty was on one too, transitioning towards a brighter future, wilder and richer in life. Our rewilding story continues beyond the bounds of this book; it continues to this day. I can't tell you how it ends – rewilding stories have no end – but in *A Sky Full of Kites* you'll learn how it started.

Life's turning points tend only to reveal themselves in hindsight, but I see now how the arrival of reintroduced kites onto Argaty came to shape my life. Had they not happened upon us at the precise moment that they did, we might have no interest in raptors, species reintroductions or rewilding today. But those birds did come, and they made so many things possible. I hope that my family played a small part in shaping a better future for them; it's the least we could do in return for all that they gave us.

Their future does look rosier than ever. In May 2023, the oldest known kite in Britain died. Named Aragon, he had been brought over from Spain as a chick in 1994 as part of the reintroduction to the Chiltern Hills. That a kite could live to such a grand old age is in itself a sign of success – it tells us that our skies are a safer place than they once were. That Aragon must have been the ancestor to some of the young British birds now being taken annually to help bolster falling Spanish populations is more wonderful still. Britain had but a handful of kites when he was released here. But in his 29 years, the population grew so much that we could help boost other countries' populations. In Aragon's story we learn all that we need to know about the success of the project to restore red kites to this small island.

In our part of Scotland we continue to bear witness to this remarkable revival. This past January, record numbers of kites were recorded flying into communal roosts across Central Scotland, Tayside and Fife. Our total of 344 dwarfed the previous year's count of 165, suggesting both that the population is growing very well and that, due to increased public scrutiny, there may have been fewer cases of persecution by gamekeepers of late. The latter point will come as a blessed relief to all who care for the birds' welfare. As you will learn in this book, the one black cloud still hanging over the restoration of red kites has been illegal killing. However, this sad chapter in Scotland's history may now be drawing to a close.

In March 2024, the Scottish Parliament passed the Wildlife and Muirburn Management (Scotland) Bill. Henceforth, any estate wishing to shoot grouse will have to apply for a licence, which could be permanently revoked if evidence suggests that, on the balance of probabilities, wildlife crime had been committed. The Bill came after evidence proved the link between raptor persecution and driven grouse shooting. For the criminals in our countryside, the noose is tightening. For our raptors, this can only be a positive thing.

One day, provided that we continue to help rather than hinder them, kites will be present throughout Scotland. We will consider them a beautiful but commonplace sight, just as our ancestors did for so many centuries. For those of us who have championed these birds for so long, that day cannot come soon enough.

Returning to *A Sky Full of Kites*, looking back on the life we've led, has proved a pleasant experience for me, like reading an old diary or flicking through a photo album. There was much in these pages that I had forgotten. I'm glad to rediscover these things, glad to remember how our rewilding project came into being. I hope you'll enjoy the story. Thank you for your interest in the book, rewilding and in Britain's most beautiful bird of prey.

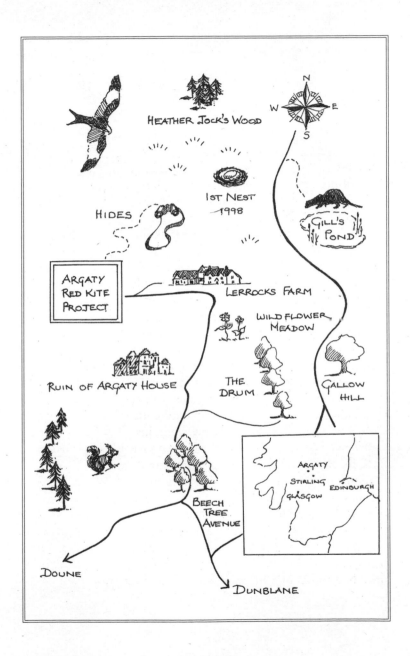

Gled

Noun

1. The common kite, *Milvus milvus*. Often used with
   the epithet greedy
2. Fig. of persons: one of a greedy disposition, one who
   preys upon his fellow-men, a rapacious person; a
   plunderer

Dictionary of the Scots Language

Foreword

I often think of the wild Scotland there once was. As the last ice age retreated, nature began to strike its own beautiful balance in myriad forms of life. Plants, mammals, fish, insects, birds, all growing, scurrying, splashing, buzzing and flapping their own beat to the same tune. A diverse orchestra of life. From the bottom of the deepest lochs to the highest mountain tops, the land was rich, alive. Wild things were unimpaired, conforming only to the rules of nature.

It's easy for us to forget that such times existed, but this was what our land was like 12,000 years ago, when it had no name and few people. But over time humans thrived and brought change, and that change has often resulted in loss. Loss of habitats, of species, but also the loss of natural rhythms and the straining of our connections with the wild world.

However, the foreword to this wonderfully uplifting book is not the place to self-flagellate. We humans, after all, are an incredible species. Admittedly, a species that seems flawed and may have lost its way, but as you'll read in these pages, instances of redemption and salvation do exist.

In today's uncertain world, we must not let loss in the natural world overshadow the gains that have been made and continue to be made. We must always celebrate how remarkable people are helping to heal the wounds that have been inflicted. Tom Bowser and his family are just that kind of people. When red kites first appeared at their Stirlingshire farm two and a half decades ago, I'm sure that no one – especially twelve-year-old Tom – realised that those dancing flames in the sky would be a symbol of change, and ultimately of hope.

I come from farming stock myself and have that perennial bond with the natural world that runs deep in families who have worked close to the land for generations. I am aware of the potential there is to prosper and to fail. Farmers, crofters and estate owners are the custodians of the land. But what does that mean in the twenty-first century? For me, a custodian of the land should be aware of the past, have both feet in the present and an eye firmly set to the future. Caring for the land is, of course, about hard work, passion and dedication, but most importantly it's about vision – the vision to perceive a future in which the balance of nature can be made more even and fairer for all that rely on what it provides, not just for humans.

As you will read, the decision made a quarter of a century ago, when Tom had his first sight of red kites in the skies over his home, was to welcome change, to nurture and celebrate life. That decision has transformed not only the trees, the fields and skies above Argaty: it has transformed the lives of the people who share that special place with these remarkable birds.

Gordon Buchanan
May 2021

Preface

As a child, the sounds I knew better than any others were those of my parents' footsteps on the stairs of our house. My mother's soft and gentle, two words that aptly describe her personality. My father's the complete opposite. He is a tall man and moves as tall men do: powerfully, with purpose.

My father was always in a hurry, always rushing from one task to another. In certain seasons he would return late more often than not. Sitting up in bed, I'd fight sleep and listen out for him so I'd know he'd made it back safely. It was in these years that I became expert at understanding his step. From the speed, I could tell whether he had finished work or was merely stopping in. From the tone, I knew a lot too. If it sounded flat he was tired, if it was light he was okay.

Often a spring in his step signified something else: good news. I was sitting on my bed one afternoon in 1996 when I heard him running up the stairs, taking them two at a time, telling us to come outside. Quickly.

It was that strange, indefinable time of the year when summer is fading but autumn has yet to truly arrive. I was twelve, about to turn thirteen, and would, I'm sure, have been very much aggrieved to be torn away from whatever occupied me that day and frog-marched out of the house.

My mother and sister were already waiting as I followed Dad downstairs. He hadn't even taken his wellies off. Crosshatched clumps of mud had fallen from the soles, leaving a trail all the way through the house.

'What is it?' Mum asked as he hurried us into our boots.

'There are red kites circling over the woods.'

Red kites? Were we really being summoned to look at people flying kites?

We followed him into the yard, stood between the orange brick buildings with their leaking gutters and mossy slate roofs, and scanned the skyline. I could make out nothing unusual.

'What . . .' I began to ask.

And then a noise filled the air. A long, shrill shriek akin to a shepherd whistling a dog, a sound so loud it would carry for miles. From somewhere in the distance there came another call. Then another sounded, and this one was nearer.

A bird soared high over the farmyard. A big bird with shining orange plumage, black and white gloved wingtips and a tail shaped like a serpent's tongue. There were more on the horizon. Suddenly the breeze kicked up and as one they reacted, twisting, looping and banking on the current. More splintering shrieks. The birds were in conversation and I was dumbstruck.

We always had pets in the house. At all times my parents maintained a squad of between five and eight dogs and at least one cat; ours being a livestock farm, there were the sheep and cattle too, even wild boar for a time. My childhood was filled with animals, but I'd never looked properly at a wild one before. Not until that moment, in the summer of 1996, when kites first arrived on Argaty.

In that moment, everything changed.

*

What those first kites were doing flying over our yard was unclear to us back then. Few people knew that the Royal Society for the Protection of Birds (RSPB) and Scottish Natural Heritage (SNH) now renamed NatureScot, had begun reintroducing chicks to the two estates immediately west of Argaty. Only when the birds skipped the border and set up camp on our farm were we admitted to that small circle of trust.

Central Scotland was, it transpired, the next link in the chain of UK-wide projects to return the gled to our skies. (The name gled, which pre-dates kite, derived from the Anglo-Saxon *glidda*, meaning 'to glide'. Kite, in common use by the 1800s, also has Germanic origins, evolving from the word *cyta*, an onomatopoeia of the birds' call.) The RSPB had first begun releasing Swedish kite chicks to Inverness-shire in the Scottish Highlands and to the Chiltern Hills in England in 1989. Leading the projects was the great Roy Dennis, the legendary ornithologist who had reintroduced sea eagles to Scotland back in 1975. A further release had started in the Midlands in 1995.

Following centuries of persecution, kites had been pushed into extinction in Scotland and England by the early 1900s, while in Wales a few clung on. At its lowest point, the number of kites there fell to ten breeding pairs, all genetically traced back to one female, but thanks to the incredible work of a small but dedicated group of protectors, those last birds survived. Although the kite population grew a little each year, they are a pack animal, loyal (as many birds are) to their place of hatching, and their numbers were too low to force them to spread back out of Wales in search of food or nesting territory. Reintroductions offered the best chance of swiftly restoring them across the rest of Britain. Given their strong attachment to their natal area, these birds would be slow to recolonise had they only been released in one spot, so for this reason they were reintroduced to various different sites, spaced at least a hundred miles apart.

Between 1996 and 2001, 103 German-hatched chicks would be imported into our area. The movements of young kites are highly unpredictable; once released from their aviaries they can go anywhere. Somehow, by some twist of fortune, these birds chose Argaty.

They chose us. That thought always puts a smile upon my face.

In those days ours was something of a road to nowhere. The only people who drove it were those who lived here. That soon came to an end. I've always wondered how the word could have spread so quickly. The internet was in its infancy then, so how did so many people learn that the birds were here? Telephone call upon telephone call is the only explanation. I like to think of it as a massive game of Chinese whispers, one birdwatcher phoning another, who then phoned another, each conversation beginning the same way: 'Have you heard the news?'

Scores of people descended upon Argaty. My abiding memory of that time is of cars abandoned at the side of the road, their doors open, the occupants nowhere to be seen, a vehicular version of the *Mary Celeste*. All was well for a time. We grew accustomed to the increased traffic and to the madcap drivers who, without warning, would slam on the brakes as a kite appeared overhead. Most of those early visitors were content to watch from a respectful distance, but there were some who gave cause for concern. These people wanted to get too close and we feared for the kites. One of the first things I learned about these great, majestic birds is that humans are the only significant threat to them. The reason they had to be brought back was that people had extirpated them, mistakenly believing that they predated livestock and gamebirds. Time passed, more kites were reintroduced to the area and the numbers roosting in our woods at night continued to grow. Unwittingly, we'd become landlords to much of the Central Scotland population. In 1998 they began to nest on the estate, too. Due to their scarcity, a kite egg was a rare and desirable thing. Although it had been illegal since 1954, egg collecting remained a very real danger. When ospreys had first returned to breed in

Britain, their eggs were regularly targeted by thieves. What would happen as more kites reached breeding age? How could their eggs be kept safe?

In a situation like this there are two options: turn people away or invite them in (albeit in a supervised way). My parents chose the latter. In partnership with the RSPB they set up a project. Each afternoon they, or one of our brilliant team of rangers, would take visitors to a viewing hide and tell the kites' story, from persecution to extirpation to eventual reintroduction; then they'd put out a small, supplementary feed – enough to top-up what the local birds found in the wild, not so much that they'd hold every bird here or make them tame – and they'd watch as the kites descended upon it. We do the same thing to this day. In my experience many farmers would just as soon die as invite members of the public onto their farm. Just as many are appalled by the thought of welcoming raptors. My parents did both of these things. I will forever be proud of them for that. Argaty wouldn't be the place it is today without the birds or the people.

At the time of writing this book it was twenty-three years since kites arrived at Argaty. That feel feels like a long time! When I first saw them quartering the sky above the farm on that long-ago afternoon, I couldn't know how they would change my life, couldn't know that many years later I would be lost and these birds would help me find my way.

Tom Bowser
Argaty
May 2021

Introduction

Summer 2017

A knock on the door.

'You there, mate?' a familiar voice called.

Mike, head ranger at the kite project, came wandering into the house.

I'd just put my daughter, Rowan, to bed. My wife, Sarah, was working late in Glasgow.

'How's it going?' I said, coming downstairs. 'Beer?'

I was surprised to see him. Although we lived two minutes apart and had been great friends since he'd come to work for my parents nearly fourteen years earlier, these days we didn't see much of each other outside of work. 'What's the point?' one of us would say whenever anyone questioned this. 'I see him all bloody day. Do you think I want to spend all evening with him, too?' At which point we'd both start laughing.

I'd barely taken a sip from my can when Mike said, 'Mate ... I've got something to tell you.'

I braced myself, fully aware of what I was about to hear. I had, I think, suspected that this moment might come for years – or if not suspected, then feared it would.

'Anna's got a new job,' he said. 'We found out today. I had to come and tell you.'

'Where?' I asked.

'Lancashire,' he said.

'When are you going?'

'August.'

I did the sums in my head. A matter of weeks. Some terrible numbness came on. But while sickness hit my stomach and my brain tried to process what it was feeling, a smile somehow manifested itself on my face.

'That's great,' I said. 'I'm happy for you.'

He surveyed me a moment.

'Honestly,' I said. 'I always thought this would happen. All your in-laws are down south. Bringing up kids with no support is tough. I get it.'

The rational part of my brain was off to a fast start. The emotional part had been left standing.

Sarah came home some hours later. Beer cans and bottles were strewn on the floor, half a bottle of whisky was gone, the Pulp album *This Is Hardcore* was blaring from the stereo. And in the midst of this carnage two drunk men were sharing old stories, laughing and trying not to cry.

Images kept racing through my head. I remembered our first conversation, the sudden breakthrough moment when we discovered a mutual love of music; nights out in the village pub, Mike, recently dumped, getting gradually more drunk, being suddenly and hilariously sick into the urinal adjacent to the one I was using; walking home from one such night, him trying to warn me about a water-filled hole in the road, then disappearing waist deep into that very hole; all the very many times I made a fool of myself in front of him and never lived it down; me serving as his best man, him serving as mine; visiting him and Anna at the hospital the night his mum passed away; him rushing to the farm on a day off to check on me after my dad came close to drowning in a boat accident; meeting his baby daughter, Edith, for the first time; Edith's first meeting with baby Rowan, born just a few months later. Memories, sweet sad memories, coalesced into one thought:

'Whatever happens in the rest of our lives, wherever we go, whatever friends we meet, nothing will be able to replace the things we went through together.'

In my mind this was a question. When I voiced it, I hoped he'd concur. But it came out as a statement, drunken, assertive, wrong. The question went unanswered and the moment was lost.

I woke at 4 a.m. with a terrible hangover, the sort that leaves you physically and mentally sick for days, and I cried and cried. Those short minutes when my rational side had taken over and I'd assured Mike that I'd known this was coming, that I understood, seemed a distant memory now. Yes, I had known this moment was coming. Yes, I understood. I understood that while the move *was* about Anna's work and being closer to family, it was also about Mike. He was no longer happy here.

In the early days of our friendship we'd often talked of Argaty and the amazing potential it had for conservation. I'd indulged wholeheartedly, little knowing the weight those conversations would carry, the expectations they'd create. I'd set myself up as someone willing and able to put big ideas into practice.

Mike had been working for my parents since 2003, when they'd decided to turn Argaty Red Kites into a full-time business rather than the weekend-only set-up it had been prior to that. His impact on Argaty had been incredible. All around the farmyard were bird boxes made from bizarre offcuts. My favourite was the old bathroom cabinet into which he'd drilled bird-sized holes.

John, our colleague at the time, used to joke that no piece of wood, metal, plastic or paper could ever be thrown out at Argaty. 'Give it to Mike, he'll make a bird box out of it,' he'd laugh.

Though the bathroom cabinet, in particular, looked ridiculous, it worked. I still smile when I see starlings disappear into it with nesting material each spring.

Then there were the steading windows, all of which were broken

long before my time. Who broke them? Why were they never
fixed? God only knows. They were broken, they had always been
broken; I hadn't given them a second thought. Mike had, though.
He removed the smashed panes and replaced them with Perspex.
In the middle of the new panes he drilled a hole big enough for
swallows to fly through but small enough to prevent magpies
following them inside. And this worked, too. Every summer the
farmyard was filled with swallows. Often I'd look through the
old, dusty buildings, most of which fell into disuse when newer,
more practical sheds were built, and marvel at the number of little
muddy nests in there.

One year I remember him drawing my attention to a little
fellow, perched on the electricity wire that runs high above the
yard. 'See his leg?' he said, pointing towards a small white ring
around the ankle joint. 'I fitted that two years ago. That swallow's
been back and forth from Africa twice since then.'

Mike had changed the place. For the better. But that cabinet
bird box rather symbolised his time with us. He was forever
working with offcuts. We never gave him any new materials. And
I'd done nothing to change that.

*

There was never any pressure on me to come back to the farm.
My parents actively encouraged me to explore all options before
settling to any one course, but in the summer of 2009 I did return.
At the time I was working as a freelance journalist and, although
I was making some progress, getting articles into national broad-
sheets like *The Scotsman* and *The Herald*, I was struggling to make
ends meet.

'Why don't you come home, help with the farm and the kites?'
Dad suggested casually. 'Go and do your writing as and when you
need to.'

It was only ever meant to be a short-term arrangement, but I found I enjoyed seeing more of my parents and learning about the kites from Mike. More than anything I loved being back at Argaty, being home. Print journalism (the only kind that held any interest for me) was on its knees back then: staff were being laid off and papers were dying a slow death, so pursuing that life would have been hard. It was online news that was growing. The farm offered a more comfortable, unthreatening alternative. The temporary fix soon became permanent, and the journalism was quietly jettisoned. Mum and Dad seemed pleased to have me there. My paternal grandmother, a wonderful and powerful woman who has since passed, was apparently delighted too. I loved her without knowing her overly well. The Bowser side of the family is big, in every sense. Big in number, big in personality. In their company the younger me – so quiet, so mousy – tended to go unnoticed. And yet my grandmother wept when she heard I'd returned home. That touched me then. It still does now. People were waiting upon my decision. That had never previously occurred to me.

I spent eight years farming with my dad, learning to work collies, shift sheep, drive cattle, fix tumbled dykes, mend broken fences, lamb ewes and all of the many other constituent parts of the job. Occasionally I'd also deliver the kite talk when Mike had days off. The idea was that I would one day take over the farm and Dad would retire. In truth I never really wanted to be a farmer, never thought I'd be good at it, never felt comfortable with the knowledge that we produced animals for slaughter. I can admit that now. I couldn't at the time. And because I couldn't I made no moves to allow Dad to step back, no plans for the future, and those big dreams Mike and I once spoke about remained just that.

Now Mike had had enough. I'd lost him. As I lay there in the silent hours of early morning, thinking of the awful night that had just passed, one moment came back again and again.

'I can't wait to see what you'll do with Argaty,' he'd said.

At first I'd mistaken his meaning, thought I'd heard some affirmation in his words.

'He knows I *will* do something special with the place,' I'd told myself. 'He knows it wasn't all talk.'

A second later my brain had caught up.

'I can't wait *around* to see what you'll do with Argaty.' That was what he'd meant. I knew that I'd feel just as he did were our roles reversed. And I hurt like hell.

Mike leaving would mark the culmination of a tough few years. I'd shredded my knee playing football, almost lost a finger in a horrifically bloody and painful farm accident, had my back give out as a result of the weakened knee. All my life I'd been fit, healthy, able to do whatsoever I wished. Now my body was like a vehicle you were afraid to drive any distance for fear it broke down on you. More recently, restored to half-fitness, I'd gone through another year's lambing. It had ended badly, with the horrible death of a lamb. Normally this wouldn't have unsettled me. Growing up in a farm environment, you see a lot of deaths. In your head you know it's sad, but eventually that sadness stops penetrating your heart. This time it had been different. I had had to euthanise the lamb. I'd never done that before, had always managed to absent myself and leave the job to someone else. There'd been nowhere to hide on this occasion. I'd come home that night deflated, flopped down on the sofa and told Sarah about my day.

'Are you sure you can do this for the rest of your life?' she'd asked.

'What do you do with a place this size if you don't farm it?' I'd replied with a helpless shrug. 'I love it. I don't want to have to sell it.'

Nobody knows me like my wife. Nobody hits the nail on the head quite as she can. Although she'd let the matter drop, her question had stayed with me, giving me no peace. Now, facing up to a future without Mike, I revisited it. Could I really do this for the rest of my life? Much of what had sustained me to that point was the company. With Mike gone and Dad talking of retirement, the future looked a very lonely place. For so long I'd been ducking big questions – Was I suited to farming? Did I want to be a meat producer? – but I couldn't hide from them forever. There would come a time, no doubt when I was carrying this place alone, when they'd become too big to avoid. What to do instead, though? I was thirty-three, with an employment record of false starts and failures. Then there was the eternal question: what to do about Argaty?

Amid all these worries and unhappy thoughts, one positive idea did come to me. When Mike left, someone would have to manage the kite project. I loved the kites and had always enjoyed running the tours. It was something I felt I did well, even if I was merely filling in on Mike's days off. At present the kites could hardly be considered a full-time job – chopping meat for them and delivering the talk took a couple of hours at most – but might it give me something to build on? Was there some way of scaling it up, making it into a career?

After so many years of indecision and inactivity, it seems unbelievable now, but within the space of a few hours I had the beginnings of a plan; an embryo at most, but more than I'd had in years.

Later that morning I found Dad in the shed, feeding the cattle. I stood beneath a shrivelled old sheaf of holly, suspended from the roof years earlier in the belief that its medicinal properties cured calves of ringworm disease. When he'd finished feeding, I asked if we could talk.

For the first time in my life I found I couldn't look him in the eye.

'I can't do it,' I said. 'The farming. It's not me.'

I wanted to say more, to explain that this wasn't a rejection of the things he and Mum had worked for, that I'd never felt pressured into farming by them. Any sense of obligation had come from within, from my desire to make them, and the rest of the family, proud. I wanted to tell him how sorry I was for wasting everyone's time. I hadn't wanted to disappoint anyone, so had kept on going. I wanted to say so much, but I couldn't. I'd have burst into tears.

And my dad – my strong, unflappable dad – did the best thing he could have done. He nodded and said, 'That's okay. Tell me what you want to do.'

Those weren't happy beginnings – losing my friend, shedding that old, ill-fitting skin . . . these things left their share of bruises – but they spawned something good. My life did indeed change. At last I did those things I once boasted I'd do. I began working to enhance the wildlife on Argaty.

As this work gathered momentum I started to think about the kites, those wonderful birds that had made all of this possible. I found I wanted to know more about them, to untangle their history and understand how they'd come to be in my life. So I began researching, following a trail back through the centuries. The information was out there, scattered across various sources. Little of it had been collated. I've sifted through it all and selected quotes that tell a story that has largely been overlooked in modern history: the story of red kites in Scotland. It's a remarkable tale. I hope you'll see just what a miracle it is that these birds are back in our skies again.

The kite story is just one of the things I want to tell you about in this book. I also wish to tell the story of Argaty Red Kites, of my family and friends, and of the work we've done to restore these raptors to our small corner of the world. By the closing pages I hope you'll know who we are and why all of this matters so much to us.

Lastly, I want to tell my own story. From those troubled days in 2017, something amazing has grown and Argaty is changing. Already we've greatly improved conditions for wildlife on our home. In the years to come we'll make them even better. In putting this down on paper I'm leaving a record of the period when all of this began – from 2017 to 2019 – a happy, sad and ultimately wonderful period of our lives.

The life I lead now is very different to the one I led when this journey began. That is all thanks to those beautiful fork-tailed raptors that first soared into our lives and claimed our hearts more than two decades ago. This book is my love letter to them.

New Beginnings

1

Argaty

He ca'd the gaislings forth to feed,
There was but sevensone o' them a',
And by them cam' the greedy gled,
And lickit up five – left him but twa.

'The Wife of Auchtermurchty', Old Scots poem
(fifteenth or sixteenth century)

Huge beech trees stand sentry at the foot of the Argaty road. More than two centuries old, their hulking arms link in the sky, forming a guard of honour, welcoming visitors in. Old woodland as an entrance to an old estate. Somehow that's always seemed fitting.

Beneath the boughs of those giant trees, the forest floor comes alive in spring. Thick carpets of bluebells stretch as far as the eye can see. Beautiful flora as an entrance to a beautiful estate. That seems fitting too.

The road winds on, burrowing deeper into this old place, further away from the outside world. On a hill to the east, another expanse of woodland comes into view. This is the Drum, the word *drum* being defined in the *Scots National Dictionary* as 'a long narrow ridge or knoll, applied to little hills, which rise as backs or ridges above the level of the adjacent ground'. (In a strange coincidence, my great-grandmother was one of the editors of the dictionary.) Giant oaks have grown in this wood. Many of the ships that sailed to battle in Trafalgar were made from them. During the Second World War the Home Guard sat atop the Drum, under the forest's shelter, keeping watch for enemy planes. The Drum is a proper wood, a wood as one imagines woods ought to be: open, dynamic, an assortment of cherry and birch, lime and Scots pine. Some of its trees are dead, others dying, but scrubby birch is growing in their place. This wood is a living creature. With no help from us, it heals itself.

To the west are vast grass parks. They sprawl, in the way you imagine the American plains do. The fields are dotted with trees. Sheep huddle beneath them on rainy days. Song thrushes potter across the grass, flit onto slumped, rotten fence posts, then fly on again. Buzzards bask in the sky overhead. From somewhere in the distance, a kite whistles.

The sad figure of Argaty House looks out onto this scene.

Once, Argaty was part of Murdoch Stewart's lands. Stewart served as Governor of Scotland in the 1420s, when King James I was held captive in England. Relishing his time in power, the governor did little to free his monarch. When at last political pressure led to James's release, he had the governor executed for treason. The Stewart land, including Argaty, was forfeited to the king. In 1478–9, James IV gifted the estate to Sir Patrick Home of Polwarth for services to the crown. He would preside here, in this strategic stronghold, keeping watch over the district for his king.

The Home family lived at Argaty for more than 400 years. In that time thousands of trees were planted across the previously barren landscape. Cromwell's invading army burned down the first Argaty House, a second was erected (then taken down again), and in 1857 the third and final version was built. The baronial sandstone mansion was once the beating heart of this place. A grand, opulent palace, it dominated the skyline.

My great-great-grandparents, Thomas and Annie Henderson, bought the house and the surrounding 1,400 acres in 1916. Thomas's arrival at Argaty is captured beautifully in the diaries of his future son-in-law, David Charles Bowser (known by most as Charlie): 'At 12 o'clock on the 11/11/16 as Mr Henderson approached the house, two doves from the Dove Cot by the Yew Tree on the front lawn were perched above the front door.'

The Hendersons were the first of four generations of my family to live in the building, although Thomas's time was short-lived. He died in 1917, a year after buying Argaty. My great-grandfather Charlie (who married the Hendersons' daughter Maisie in 1922) was the first of our family to truly serve as laird of the estate. He and Maisie modernised agricultural practices. They built houses. (Much of the Argaty that I know today was born in that era.) They also raised three children: Hubert, David (my grandfather) and Eleanor.

The mansion served as a maternity hospital during the Second World War. Pregnant women were whisked out of Glasgow, away from enemy bombs, to this quiet corner of the countryside where they could give birth in relative peace. The babies slept in dresser drawers.

After Hubert's untimely death in the war, my grandfather fell heir to the estate. He and his wife, Judy – the wonderful, powerful woman who wept those unexpected tears when I returned to the farm – would raise five children in Argaty House. My father and his sisters would be the last generation of the family to live there.

Although it was truly the most marvellous building, it was also a massive financial drain, always taking money, never giving back. When my parents moved from London to take over the estate, the mansion was sold. The house I grew up in, Lerrocks, was the old farm manager's dwelling.

There is nothing beautiful about Argaty House now. In April

2011, an electrical fault started a fire which gutted it. The external walls are all that remain.

Beyond the ruined mansion lies a walled garden, which was still operational in my childhood. I can clearly remember the lush flowerbeds, the endless rows of vegetables, the cloying, stomach-churning odour of tomatoes in the greenhouse. Once, enough fruit and vegetables were produced within those walls to feed every man, woman and child on the estate. From foresters to gamekeepers, farm managers to farmhands, nannies to cooks to other domestic staff, so many people earned a living here. An estate was a self-sustaining community then. Advancing mechanisation brought that era to a close. With only a handful of people working the land, this garden was rendered redundant. The potting sheds have since been converted into a house, the vegetable plots and flower-beds have gone to grass. I am one of the few people still living who remembers what it used to look like.

As we pass the garden the road begins to climb, towards the farm, towards the kites. Arrive in the afternoon and you're likely to see the birds swooping for meat in the fields behind the steading. Little squads of them descend, screaming their battle cry, grabbing the food and going. This is their land, their territory, their ideal habitat.

By now we've left the gentle, picturesque parts behind and entered another country altogether. Argaty sits on the Braes of Doune, ten miles north of Stirling, the one-time capital of Scotland. Our part of the world is often called the Gateway to the Highlands. This is where the civilised Scottish Lowlands end and the weather-worn, calloused uplands begin. Argaty is a mix of the two: a meeting of these very different landscapes and cultures. The half just described is the civilised part. The upper end is any-thing but.

The name Argaty – pronounced *'Ar-gut-ee* – translates as high, windy place, and the name is apt. These hills are lands of sphagnum bogs, rampant gorse, endless rushes. Nothing can be cultivated here. No machine could handle the terrain. Agriculturally, it is useful for one thing only: rough grazing by hardy breeds of sheep and cattle.

The outlaw Heather Jock, immortalised in the following ballad, once hid out here, rustling livestock, spearing fish in the river, wreaking havoc:

> Heather Jock's noo awa',
> Heather Jock's noo awa';
> The muircock* noo may crousely craw
> Since Heather Jock's noo awa'.

> Heather Jock was stark and grim,
> Faucht wi' a' would fecht wi' him;
> Swamp and supple, sharp and thin,
> Fine for gaun against the win';
> Tawnie face and tousie hair,
> In his cleading unco bare,
> Curs'd and swore whene'er he spoke
> Nane could equal Heather Jock.

*muircock: the male of the red grouse, *Lagopus scoticus*

The ballad was written after a judge deported Jock to the penal colony in Botany Bay, Australia, in the mid-1800s. This was the second attempt to banish him. Due for deportation some time earlier, he'd escaped custody and had to be recaptured. Between times, Jock's mother, ashamed at her criminal son, hanged herself. Legend has it that he cut her down purely so that he could steal the rope. Truly, none could equal Heather Jock.

Nowadays we'd call Jock's lands 'Less Favoured Areas', a term coined by the European Union to describe places where farming production is hindered by natural handicaps. In former times, however, this part of the estate was very much favoured. The reason? Shooting.

These moors were once covered in heather, an ideal habitat for partridge and grouse. No doubt that is why my great-great-grandparents decided to part with some of a small fortune made importing tea from India and buy the estate. Purchasing land was something of a trend in that era. In 1848, Queen Victoria and Prince Albert began leasing Balmoral estate in Aberdeenshire. They were said to be entranced by the Highlands, and the hunting and fishing opportunities available there. Four years later, they bought it. Many estates across Scotland changed hands over the following decades as the nouveau riche sought ways to emulate the monarchy.

You'll find neither grouse nor partridge crawing on our muirs now. As sheep numbers were increased, the heather cover reduced and the birds' habitat was destroyed. The rabbits, which were also hunted here, have all but disappeared too, falling victim first to myxomatosis and then haemorrhagic disease. This part of the estate has changed so much.

The same could be said for much of these uplands. Old maps show a line of farms stretching for miles and miles across the hills. Most of them are gone now. The ground was too poor, the life too hard. Proof of that can be found in the tale of old Mrs Bell of neighbouring Calzie Bohalzie, who died in the long winter of 1947. Stranded on their hilltop, all roads blocked, her family had no option but to bury her in the snow and wait for the thaw. Those were hard people living in hard times – harder than most could stomach now. Argaty is one of the last estates standing, a remnant of that era.

This is my home, an estate of two identities: the polite public face and the feral, untameable heart. There's so much wildlife to enjoy here. Kites, kestrels and barn owls search for voles on the moorland. In the hedgerows, songbirds dart to and fro, while out in the lush green parks red and roe deer roam wild. By day, red squirrels leap through the canopy of our broadleaved woods; by night, foxes, badgers and pine martens take over.

There's history to uncover, too. It's visible and it's palpable. Study a map and you'll find the Judge's Cairn, Heather Jock's Wood, the ominous-sounding Gallowhill. Every landmark has a name and every name is a link to the past. Many of our woods are called after somone who was born or died here. Generation upon generation tried to leave something of themselves behind, something that would remain long after they'd gone. Even now, we still do the same. One mild, starry night back in 2009, sitting on a silage stack near our red kite hide, I asked Sarah to marry me. I thought hard about her wedding present. I wanted to give her something that represented the qualities she has, something strong and beautiful that enriched the world around it. I chose a tree. It's planted in view of the silage stack, thirty yards from the hide. When our daughter was born, we thought of that tree and named her Rowan.

*

My sister and I were the fifth generation of the family to live on Argaty. Rowan is the sixth. Many generations of kites have grown up here, too. Ever since that day when they first appeared over the farm, we've had them here.

In the spring of 1997, a year after the first kites were reintroduced to the area, a pair of one-year-old birds were frequently seen flying in tandem over the woods beside my parents' house. Though they don't usually breed that young, they'll often pair up and claim a

territory in advance. Trackers had been fitted to all of the reintroduced birds so that their movements could be monitored.
The RSPB's head of conservation, Duncan Orr-Ewing, who had
previously worked under Roy Dennis on the 1989 Black Isle
reintroduction, managed the Central Scotland project. He and
his project officer, Lorcan O'Toole, came here daily to check on
the birds' progress and provide them with a supplementary feed
of carrion. Often they'd knock on the door and give us an update.
Sometimes we went out with them to watch the pair in flight.
Throughout the summer of 1997 those birds barely left Argaty.
It seemed we might be on the verge of making history. If they
remained faithful to that spot, they would probably nest there the
following year.

Soon Duncan and Lorcan travelled to Germany, collected more
chicks from Sachsen-Anhalt and flew them back to Scotland, ready
to release later that summer. The birds were taken when they were
between four and five weeks old. By that stage they had enough
feathers to stay warm without their mothers' brooding and were
capable of walking towards meat and feeding themselves. They
travelled in cages in the aeroplane hold, packed carefully alongside the other travellers' luggage. As had happened the previous
year, their next few weeks were spent in large aviaries on estates
near Doune.

Thanks to our growing friendship with the RSPB team we
secured an invitation to see the birds in their cages. My main
memory of this is of sweat, pain and exhaustion. Some weeks
earlier I had broken my leg in an accident at school. Come the
time of our visit to the kites, I was still in plaster. The aviaries were
hidden deep in the woods, away from prying eyes. I remember
my parents and sister gamely trekking off into the undergrowth,
leaving me, half mummified and hating the world, trying to follow
on crutches. (In the end, I gave up and sulked my way back to the

car). A few weeks later, when the birds were two months old, the age they fledge nests in the wild, they were released.

I was too young to appreciate it at the time, but the year 1997 was a time of change. New Labour had swept to power, Scotland and Wales voted in favour of devolution, the Scottish football team even qualified for a major tournament . . . All around Britain things were happening. As Kevin Duffy, who would take over as the RSPB's red kite officer after Lorcan returned to Ireland, recently told me: 'The nineties were a really optimistic period. In recent years more sea eagles had been reintroduced to the west coast of Scotland, kites were now back in the Chilterns, the Midlands, the Scottish Highlands and Central Scotland, and there was talk of officially reintroducing beavers next . . . though that took decades to actually happen. The attitude was very much: "If a native animal was persecuted to extinction and you have the power to bring it back, why wouldn't you do it?" Everything seemed possible. There was so much funding available for conservation back then, sums that would be unthinkable now and will probably never be seen again. For anyone interested in wildlife, it was an exciting time.'

The continued success of the Central Scotland kites only added to this. In 1998, as we had hoped, red kite chicks arrived at Argaty. Again, my family were fortunate. We were allowed to visit the nest when the three chicks were tagged and leg-ringed. Lorcan climbed the tree and lowered the chicks down in a bag. As Duncan unzipped it, a blaze of scorching colours met my eyes. Flaming orange. Deep, burnt brown. I'd never witnessed anything so splendid. I've seen and handled many young kites since that day, but my first impression remains my favourite. On 6 July, the birds fledged. A few days later a further two flew a nest on Doune Lodge, our neighbouring estate. Though it's impossible to say with any certainty exactly when the last nesting pairs of kites were killed off in this part of Scotland, I've found no records of them beyond the 1880s. As far as we know, these five young kites were the first chicks to fledge in this area for over a century.

The Argaty pair remained loyal to their nest site throughout that winter. Their chicks hung around too, and this had something of a magnetic effect. Kites are social creatures and soon other birds imported from Germany in the previous two years arrived on the farm. This small marauding group spent the days searching for food together before disappearing into the trees at night to sleep. Now we not only had a nest but a communal roost too. In 1999, another pair formed and that year we had two successful nests.

I've often wondered what it was about Argaty that attracted the birds. Why were we the lucky ones? Reading back through old books and RSPB nesting reports, I started to understand.

From remains found in a cave hyena lair on Wales's Gower Peninsula we know that kites have been resident in Britain for at least 120,000 years. In his book *The Kite's Tale*, Roger Lovegrove surmises that the post-glacial landscape would have limited their numbers. Kites like open countryside. They spend most of their lives on the wing, using their formidable eyesight to seek out carrion and small prey. Large stretches of woodland are of little use to them. With too many trees, it's hard to even see carrion or prey, far less squeeze a five-foot wingspan between close tangles of branches to grab it from the forest floor. While wetlands and the clearings created by large herbivores would have provided some space to hunt and scavenge in, the trees that covered so much of our lands in that era must have proved difficult to navigate. From around 2500 BC humans began clearing woodland to make way for agriculture. This, says Lovegrove, provided these birds with exactly what they needed: open space, clear sightlines and flight-paths. Thanks to our ancestors, kite populations grew. By the Middle Ages they were thought to be our most common raptor, with large numbers found scavenging in the growing towns and cities where refuse and animal carcasses filled the dirty streets. (Until sanitation eventually improved, kites, ravens and herds of pigs provided a valuable service, feeding on much of the waste and keeping urban settlements at least partially clean.)

Meanwhile in the rapidly changing countryside their populations continued to grow as well. Argaty may be a fairly accurate modern representation of the rural landscape created all those centuries ago. Our wide, barren moors make perfect hunting grounds; our great old woods are an added attraction. Kites favour mature trees for nesting; those first pairs built their nests in an oak and a Scots pine respectively. These slow-growing species must have stood for hundreds of years to reach maturity. That tells you a lot about the people who lived and died during their lifetimes. On Argaty, trees have always been valued. Though less the case now, in the past Britain's gentry routinely engaged in public service. In this way they gave something back to society. Both my great-grandfather Charlie and my grandfather David were tree men. Both served as forestry commissioners and maintained a much larger acreage of woodland on their land than was common elsewhere. For my great-grandfather, in particular, the loss of healthy woodland was a source of devastation.

In a letter to his mother-in-law, dated 1927, he wrote of a great storm that had laid siege to Argaty:

> The view from the garden of perhaps 8 or 10 upturned
> roots and broken dykes is depressing. But in the Drum itself
> it is difficult to tell you of it, for it is heart-breaking . . . In
> some places it seems as if the wind came straight out of
> Heaven and blew trees down in all directions.

Once that storm had abated, he sent his foresters out to count each fallen tree so that every one could be replaced. That same year, my grandfather, at just one year old, planted his first tree, a scarlet oak, and a shared passion was passed down from father to son.

Of course, none of my forebears managed their woodland with kites in mind. As far as anyone in the early to mid-twentieth century knew, the birds had gone for good. Still, it was the decisions made in their time that provided a home for kites in the late 1990s. As unintended consequences go, this was a happy one. Argaty became the kites' hub, their centre of operations. In these trees their population has grown.

2

Into the Woods

If charnel-houses* and our graves must send
Those that we bury back, our monuments
Shall be the maws* of kites.

William Shakespeare, *Macbeth*, 1606

In recent times Argaty's trees have become home to something else. Somewhere out there resides another animal that was absent for most of my life. One morning in August 2017 I strode out to find it.

Though it was supposed to be late summer, you couldn't tell. The weather was grey, grey, grey. Most of the year had been the same. We'd had just one dry spell, in May, during the lambing season, when my farming life had unravelled. Now August had arrived and my mood mirrored the miserable weather. Later that month Mike was leaving.

After the drunken, tear-filled night when he told me his decision we'd grown somewhat awkward with each other. On the horizon he had the stresses of the move and of finding a new job. Meanwhile I was battling feelings of abandonment that I knew were unjustified. When you anchor yourself to a place, you have to

*charnel house: where the bodies and bones of dead people are stored; maw: the mouth, or stomach of an animal, or of a greedy person

accept that people will leave your life. Some go temporarily, others forever. I've never been angry about that. It is, after all, my choice to remain. Nobody has ever forced my hand. I get sad sometimes, though. Sometimes I just wish everyone was made like I am, made to stay where they are.

Mike could have done without all of that, I'm sure. Tensions were probably exacerbated by my decision to abandon farming and begin this search. Were it successful, Argaty would take another leap into the world of conservation. Just as he was leaving. I knew that looked like a revenge mission, something akin to a jilted lover getting a new partner and parading them in front of their ex. The truth was quite different. I really did want to do something worthwhile with my life, and conservation seemed a natural extension of the work we were doing with the kites. That wasn't my main motivation, however. My main motivation was Mike. I wanted him to change his mind and stay. I knew he couldn't and wouldn't, but that's what I wanted.

So, that August morning, into the woods I went.

I picked the nearest trees, a Norway spruce plantation that lay only a hundred yards from our visitor centre and car park. Inside the wood the air was dead. No wind cut through, no light shone in. The canopy had closed many years before and it had swallowed everything. The forest floor was a land of shadows, a sodden mat of needles, mud and puddles that would never drain. Magpies were rattling somewhere above. How many there were, one for sorrow or two for joy, I couldn't tell. They sounded distant, like something heard through thickened glass. The world outside could have been miles away.

Spruce cones littered the ground. I inspected them, twirled them in my hands, threw them aside. Each was disappointingly whole. I didn't want whole cones.

Reasoning that perhaps I was too close to the edge of the wood, I ventured deeper in. The trees closed rank, the atmosphere became claustrophobic. Every move you make sounds massive when you're the only thing moving.

Soon I saw it. There at the base of a tree was a cone and its bottom end was stripped to the core. All that remained was a little brown husk with a ragged top. It looked like a miniature palm tree.

Red squirrels. Hard to believe in that silent, gloomy place, but they were there. I couldn't see them, or hear them, but somewhere in the dark towers above, squirrels were watching me. This ravaged remnant of a cone was proof. Squirrels have a distinctive way of eating them, quite different from mice. Where mice make a neat job, squirrels destroy, spinning the cone in their paws, pulling the external scales off and discarding them, ravenously seeking out the seeds in the middle. This was it, what I'd hoped to find, and it would mark the first step on a new path for me.

In the past week my parents had begun looking for a contractor to manage our livestock. Once they identified the right candidate Dad would retire and my agricultural career would come to an end too. I had a small firewood business, set up some years earlier, which helped generate income during the winter months. I'd keep that going and take on the running of the kite project too. Both of these jobs were highly seasonal, however. If the weather was poor I might sell more wood, but we'd scarcely have any kite visitors. If the weather was fine, wood sales would drop as visitor numbers increased. Either way, I'd only ever be working one part-time job. To justify a wage I needed something else. Seeking out squirrels was the first possibility that occurred. Find them and who knew what conservation opportunities might open up for us?

That morning I'd built a feeder box – the first of many that I'd make in the coming months as I began a wider search for squirrels on the estate. In most ways these resemble a bird nest box. The difference is that the former has a Perspex front and a hinged lid. Squirrels see the nuts through the Perspex and lift the lid to get inside.

I screwed the box to one tree. As the metal bore into the trunk, oily resin glooped onto my fingers. On a tree opposite the feeder I fixed a trail camera – a handy device used by conservationists and hunters alike to locate animals. Whenever something moves, the camera takes a photograph. If, as I hoped, red squirrels found the nuts, I'd have a picture of them.

Standing back to inspect my work a feeling of immense excitement washed over me. This felt big, like the start of something.

*

Growing up, the very notion of looking for red squirrels on Argaty would have been unthinkable. In the eighteen years before leaving for university I had seen just two of them. Grey squirrels dominated this area. That remained the case until 2012 when Mike signed us up to the Forestry Grant Scheme – a five-year project to trap and cull greys in target areas of the Scottish countryside. Squirrel traps are mesh boxes, around two feet long, half a foot wide. At one end is a door, at the other a pressure pad. The two are linked by a long, thin rod that props the door open. You put peanuts or maize at the pad end. If something enters the box, enticed by the food, it steps on the pad and the door closes, leaving it trapped but physically unharmed. You check the boxes at least twice a day. If you've caught a grey squirrel, you shoot it; if something else is in there, you set it free.

Having written a dissertation on the impact of grey squirrels upon reds, Mike was an expert. He knew that the only way of saving the latter in Britain was to remove the former. Over that five-year period he would put the theory into practice, killing 150 greys here. Be it grey squirrels, mink or deer, culling for conservation sticks in the craw. It seems such a contradiction. Our ancestors eliminated so many top predators. They introduced incompatible, non-native species. Had they acted differently, culling wouldn't be necessary; we could leave nature to get on with it. Those people messed it up, wrecked the food chain, and we're still living with those mistakes. It's possible that we always will.

The introduction of grey squirrels from North America in the 1870s has been a particular disaster. As Polly Pullar and Neil McIntyre highlight in their book, *The Red Squirrel: A Future in the Forest*, red squirrels had already suffered greatly at the hands of man. Over the centuries we had steadily reduced our woodland cover, felling trees for fuel, wars and to create space for agriculture. In doing so we destroyed the vast, connected habitats that squirrels relied on. We also hunted them mercilessly, blaming them for damaging forestry by stripping bark and eating acorns and nuts. Unbelievably these hunts continued well into the twentieth century. From James Ritchie's *The Influence of Man on Animal Life in Scotland,* we learn that 'in the woods of eastern Ross-shire, part of Sutherland and that portion of Inverness north of the Caledonian Canal,' 60,450 were killed by the Highland Squirrel Club between 1902 and 1917.

Already in steep decline, red squirrel populations plummeted in Britain with the arrival of grey squirrels. Bigger and more aggressive than our native reds, greys are also more adaptable eaters. Where a red must wait for an acorn to ripen before it can digest it, greys can eat them green. The food is gone before the reds have a chance at it. A further devastating impact of the grey squirrel saga is the squirrel pox virus, discovered in England in the 1980s. Carried by greys, which are immune to it, the virus is fatal to reds. It causes sores, lesions, disorientation, and loss of energy and fur. Within two weeks, the red will die. Terrifyingly, the greys have brought the virus north, deep into Scotland. It was recently located near Stirling, fifteen miles from Argaty, in a country park filled with greys.

Grey squirrels have already taken over most of England. Now reds remain in just a handful of places south of the border. They may well die out there. Much of Lowland and Central Scotland has been lost too, as the greys spread towards the red-only Highlands. It's worth stressing that grey squirrels aren't to blame for thriving on foreign soil. People brought them here. We are the real culprits.

Nothing gives me less pleasure than the thought of killing a wild animal, but in order to redress the imbalances that we have created we must sometimes be willing to pull the trigger. Some might baulk at that, some actively rally against it, always on animal welfare grounds, but we face a stark choice: should we cull grey squirrels as quickly and humanely as we can, or do nothing? By opting for inaction we abandon reds, allowing them either to catch the pox and die in slow agony, or else be forced from their homes and left to starve. Surely this is the greater animal welfare violation.

As the old saying goes, nature abhors a vacuum and the upside to the Argaty grey squirrel culls is that reds have come flooding back. Red squirrel charities frequently cite Mike's work as an example of the effectiveness of grey squirrel control. Neighbouring estates are also culling and now, for the first time in decades, reds are appearing in local towns and villages. Initially I had no idea quite how great their revival had been. They may have been out there, growing in number, but I was none the wiser. Only in recent times had I realised.

For years I'd been feeding birds in my garden, attracting the usual mix of tits, chaffinches and robins. The peanuts in my mesh feeder had gone down slowly and steadily each week. Then, back in the spring of 2017, something strange had happened. The nuts had begun to vanish. In the morning I would top them up, by lunch they'd be gone. It had made no sense to me. The number of birds hadn't increased and, in any case, they usually took days to finish the food. Inspecting the feeder, I'd noticed that one of the mesh squares was misshapen; the wires had been stretched. What on earth had caused that? As I'd replaced it on the branch, a handful of whole nuts had fallen out and landed upon the lawn. The birds had come in and grabbed them as soon as I'd stepped away.

Several days later the gap had grown wider. One wire had come loose and spiked my hand when I'd begun to fill the feeder. I'd stared at it, perplexed.

Days later I would see the culprit. Standing at my bedroom window, enjoying a morning coffee, a red squirrel had run across the top of the garden fence. Lowering itself headfirst, it wrenched another two wires apart, spilling yet more food, and bounded off with a nut in its mouth. In the next half-hour the little vandal returned again and again. By the time it had finished, the feeder looked like an exploded porcupine, wires jutting out in every direction.

As we'd moved deeper into spring more squirrels had come. Sometimes I saw two or three running around, chasing each other up and down the fence and nearby trees, squabbling over the nuts. Often I'd lift Rowan up and we'd watch the scene unfolding together.

Later, at the start of that long, dismally wet summer I received a very kind donation of £100 to spend on conservation from a lovely American couple, Lisa and Paul Williams, who had visited the kite centre. Mike suggested we buy a trail camera to monitor birds' nests. I duly took this advice, though by that point all of the chicks had fledged and the camera had sat unopened in its box for several weeks. Every time I looked at it I felt guilty, like the money had been wasted.

We had squirrels on Argaty, I had a camera, yet I hadn't connected the two – not for some time, anyway. But then August came, and Mike was leaving, and craving both a means of making him stay and of changing my life an idea presented itself. Find red squirrels. Feed them. Set up a viewing hide. Squirrel tours in the morning, kites in the afternoon. A decent sort of life by anyone's reckoning.

Walking back through the trees later that afternoon, I buzzed with a sort of nervous excitement. What would be on the camera? The answer, disappointingly, was nothing. The peanuts lay untouched in the box. It was the same the next day and the day after. Soon the limitations of this plan became apparent to me. I had no idea what I was doing. I'd only found one chewed cone and it could have been old. How long do cones take to biodegrade? Was this monoculture wood really the place to look? It was hardly buzzing with wildlife. The only reason I'd chosen this over the many more attractive woods on Argaty was that it was close to the kite centre. If we ever wanted to build a viewing hide, it seemed a convenient spot to which people could walk.

For want of a better plan, I stuck with the plantation, adopting trial and error as my new modus operandi. Though I knew little about the animals I sought, I reasoned that I could learn a lot through my failures. Hoping that in better light the squirrels might see it, I moved the box to a tree at the edge of a clearing. Nothing came. I laid trails of nuts for squirrels to follow, leading all the way to the box. The birds were very grateful for this. They took them before even the fastest squirrel could have found them.

Days passed. No success. I became more and more convinced that the squirrels had long since vacated this wood, that I'd have to search further afield. In a final roll of the dice I abandoned the peanut boxes and turned to a trusted ally, my old wrecked bird feeder. Ducking and weaving between branches, I found the darkest, densest part of the wood, the place where the squirrels might feel safest. It was there that I hung my old friend.

And that day the first squirrels came.

3

Opportunities

Arran Castle, 15th September 1779 Premiums for
   destroying birds of prey in Arran:-
An eagle 7s. 6d.
For the nest of an eagle 10s. 6d.
A game hawk 2s. 6d.
For the nest of a game hawk, the young ones alive 10s.
   6d.
A common kite 1s. 0d.
For the nest of a common kite 2s. 6d.
For a cat shot on the muir 1s. 0d.

Robert Alexander Forbes, *Gaelic Names of Beasts*
(Mammalia), Birds, Fishes, Insects, Reptiles, etc.

History has a strange way of repeating itself. When I began my
search for the squirrels I was, in a sense, repeating what my parents
had done at the start of the kite project. If I was going in blindly,
trusting to fate that the plan would work, so were they.

In the hills above the farm, the wild lands where the whin bushes
sprawl and the wind is always a gale, where the lapwing wanders
and the curlew calls, the remains of a wooden shelter can be found.
It collapsed a few years ago. One day it was standing, the next it
was down, a little bit of history rotting into the ground. This was
our original hide, where visitors first came to watch the kites.

One night, soon after I'd begun my squirrel hunt, with Rowan asleep upstairs, Mum and Sarah both out, Dad and I lit a fire, poured a dram and reminisced about those days.

By the early 2000s we had a few pairs nesting on the farm and a growing number roosting here through the winter.

'There was a real opportunity,' Dad told me. 'I knew it must be possible to entice the kites down for food. When they first arrived on the farm, the RSPB used to come up and put meat out for them. Because the birds had been taken from the nest they had nobody to show them how to feed, like the parents would do in their first month in the wild. As the red kite officer, it was Lorcan O'Toole's job to replicate that.'

There were precedents for feeding adult birds, too. At Gigrin Farm, near Rhayader in mid-Wales, a feeding station had been established in 1992. For over a century, conservationists had been working to protect the last Welsh kites, distributing instructions for their preservation to gamekeepers, paying landowners to keep nests safe from egg collectors and persecutors, even tasking the British armed forces to guard nest trees. Now, thanks to this and to the supplementary feeding at Gigrin, the kite population was rising. A feeding station in Central Scotland might benefit the kites every bit as much as the Welsh one had.

Dad continued: 'When he took over from Lorcan, Kevin Duffy tried feeding over on Doune Lodge and found that some kites would come down. He felt that Argaty would be a better site for it, though, and gave Mum and me a presentation about Gigrin one night. We went down to Wales to see it for ourselves and it really opened our eyes to the possibilities.'

'Weren't you nervous about bringing all of those people onto the farm?' I asked.

'A little,' he said. 'Maybe Mum was a bit more hesitant than I was. There's always that fear that you get overrun, that the place turns into a zoo. But people were coming anyway, so it was better to try and accommodate them.'

'And you knew it could be done? That you could feed the birds.'

'Absolutely.'

'And how did you get on when you first started? Did they take to it straight away?'

A smile flashed across Dad's face. 'It was a fucking disaster.'

Through a contact at the Farmers' Market in Stirling, my parents acquired a vanload of breasted pheasants. One day Dad disappeared off into the hills above the farm and threw a few out where he'd last seen kites hovering.

'Absolute disaster,' he repeated. 'The kites – these bloody big birds that were supposed to come and take everyone's lambs, hens and pheasants – just looked at the meat. I tried a few different spots, but the result was the same. Their beaks and talons weren't strong enough to rip into food of that size. The only time they ever came down was when a buzzard had been in and torn it to shreds.'

*

Dad's early attempts, trailing off across the hill with a stinking bag of half-chopped birds, wondering why the kites wouldn't come for such a delightful treat, seems hilariously hapless. But there's a serious side to the tale, too.

Red kites have been persecuted for centuries because people see them as voracious, greedy birds. As a society Britain has had a difficult relationship with both scavengers and predators. (It is interesting, at the very least, that the alternative definition of 'predator' in the *Oxford Dictionary* is 'a person who ruthlessly exploits others'.) We've tended to sensationalise these animals, make them bigger, stronger and more capable than they are.

Kites are scavengers by nature, and the reality of their behaviour – that they will feed on any dead meat they can find – may explain some of our hostility towards them. (Even now, in our apparently enlightened age, vultures in Europe are sometimes viewed with disgust because they feed on meat in oft-advanced stages of decay.) Centuries ago, when dead animals littered city streets and battles between neighbouring nations were commonplace, one can only imagine how much carrion was to be found.

An interesting reference to this relates to the Battle of Killiecrankie, fought near Blair Atholl in Perthshire in 1689. John Graham of Claverhouse led the Jacobite army into battle against the government's troops. Halyburton of Pitcur, who also fought for the Jacobites, died in combat and Claverhouse succumbed to his wounds shortly after. Robert Burns's 'Killiecrankie', written in 1789/1790, commemorates this:

> The bauld Pitcur fell in a furr*,
> And Clavers gat a clankie, o,
> Or I had fed an Athole gled,
> On the braes o Killiecrankie, o!

> *furr: furrow

The image of a kite feasting on a corpse is hardly pleasant. In those times empathy towards wildlife was a largely unheard-of concept, and although the birds were merely exploiting the opportunities provided by us, we might easily have reached the conclusion that they were indeed 'ruthlessly exploiting others'.

Aside from scavenging, kites will also take small prey. By the end of the Middle Ages, when their populations were at their highest, they must have caused people difficulties. In the countryside free-range hens and geese were kept. Although kites could not kill anything so large as an adult bird, they would have been partial to their chicks. In that era many people lived in extreme poverty and squalor. Disease was rife, and if the harvest was poor they starved. Ducklings, goslings and rabbits (introduced by the Normans in the early twelfth century) were a vital food source, and it is little wonder that kites were reviled for predating their young.

Prompted by his subjects' suffering, in 1457 James II of Scotland declared that kites should be killed wherever they were found. After bad harvests and subsequent food shortages, the Tudors followed suit, passing vermin acts in 1532 and 1566. Citizens were required by law to kill any of the many creatures that appeared on their monarch's list of vermin. Church wardens ran the scheme, paying a penny for every kite head brought to them – and more for some other animals. Records still exist for several counties, particularly in the south-west. They paint a very grim picture. From Lovegrove's *The Kite's Tale* we learn that from the seventeenth to the mid-nineteenth century, when the law was finally repealed, over 10,000 bounties are known to have been paid for kite heads in Cornwall, Cambridgeshire, Cheshire, Kent, Devon, Somerset and Gloucestershire alone.

In Scotland the situation may have been different. James II died three years after declaring war upon the kites, and there is no evidence that his orders were ever acted out. The old *Statistical Accounts of Scotland*, written in the late 1700s, state that kites were common in most parishes. Up to that point, we can only infer that levels of killing in this country were not significant.

This changed, first with the Highland Clearances, when tenant crofters were cleared from the land to make way for large-scale sheep farming, and then with the growth of game shooting in the 1800s. From this point forward we see increasing examples of long-held antipathy towards predators turning to coordinated, devastating action, as gamekeepers and landowners moved to eliminate any animal that threatened, or was perceived to threaten, sheep or game. Red kites were on the road to extirpation.

Between 1776 and 1786, 2,520 hawks and kites were killed through a sheep, game and poultry preservation scheme in the Aberdeenshire parishes of Braemar, Crathie, Glenmuick, Tullich and Glengarden. (Seventy eagles, 634 foxes, forty-four wild cats, fifty-seven pole cats and 1,347 ravens and hooded crows were also killed.) On Dunira estate in Perthshire, owned by the politician Henry Dundas, thirty kites were killed in 1809; meanwhile on Buccleuch estate in the Scottish Borders, a further fifty-eight were dispatched between 1819 and 1827.

By the time that the new *Statistical Accounts of Scotland* were written in the 1840s, gamekeepers had eliminated kites in many parishes. We can't say how many were shot, trapped or poisoned in each area, most estate records being frustratingly hard to find, but 275 were known to have been killed between 1837 and 1840 on Glengarry estate, Inverness-shire, and keepers would later bag a further 105 in 1875 on the Callander Hills just north of Argaty (now known as the Braes of Doune). If these figures are anything to go by, we can draw only one conclusion: across Scotland, the death toll must have been enormous. Kites slipped silently from our skylines, county by county, year by year. By the end of the century they were extinct in most of Scotland. Some isolated sightings do exist into the 1900s. A pair nested in Spean Bridge, Inverness-shire, in 1902, but a collector stole the eggs. Another pair was recorded in Glengarry in 1917. Given the location, it can't have been long before the keepers caught them up. Beyond that date, I've found nothing. Some last survivors may have lingered on, but most, it seems, were gone.

It's perhaps worth pausing for a moment to consider the magnitude of the 'vermin' culls, the mind-boggling extent of the killing. Today in the *whole* of Central Scotland we have slightly over a hundred pairs of kites and we consider this a thriving population, a cause for celebration. Yet in former times a keeper on *just one estate* might kill close to half this number in a single year. Those dreadful statistics reveal the full tragic extent of what we once had, and what we have lost.

Possibly the saddest part of the red kite's sorry history is that it is not a voracious killer. In Dad's breasted pheasant story we see the real truth: even as scavengers, kites are fairly feeble. They couldn't even take a dead adult pheasant, far less a live one. That people in pre-Victorian times disliked the kites for taking small prey was unsurprising. That kites' scavenging may have caused some disgust is understandable too. These were not the reasons that led to their destruction, though. We killed them because we overestimated their ability to kill. Erroneous information wiped out the red kite.

*

When his first feeding attempt failed, Dad realised that a different tactic was needed.

'I started chopping the meat up,' he said. 'Having diced it to stewing steak-sized cubes, I went off again into the hills. I put the food out and retreated to a safe distance. And lo and behold, down the kites came!'

'It must have been something, seeing that for the first time,' I said.

'Yeah,' he said. 'It really was.'

For my parents, the kites had arrived at just the right time. In the late 1980s and early 1990s, the neurodegenerative disease Bovine Spongiform Encephalopathy (BSE, more commonly known as 'mad cow disease') had hit British cattle. In 1995 three people died after eating infected meat. The European Commission responded by banning the export of British beef, the government introduced a scheme to withdraw cows over the age of thirty months from the food chain, and the sale of beef on the bone was stopped. It was a terrifying time for consumers and a nightmare for livestock farmers. Already struggling to get a fair price for our produce, thanks to the supermarkets and competition from overseas, the BSE outbreak hit families like ours hard.

'We needed something,' Dad said. 'A little bit of luck. And the kites came.'

Using offcuts from our local sawmill, he and John, our former stock man, built a hide and began inviting RSPB members' groups to come and see the birds feed.

'I remember sitting in there with a group, looking up at these specks above the skyline,' said Dad. 'They were miles away, much further than the distances we watch them from now, but people were blown away just to see them.'

As he recalled this, his gaze became distant, as if at that moment he was staring at a hillside and a sky full of soaring kites.

'The hide wasn't exactly the Ritz,' Mum would later tell me. 'You had to walk through a pretty boggy, soggy approach and then wait, hoping that the kites would make an appearance. I used to take a large bar of chocolate to cheer things up, just in case the birds were playing shy! Afterward we went back to the farm and gave folk a cup of tea. It was rough and ready, but great fun.'

As time went on, the birds grew increasingly confident in the presence of onlookers. My parents were then able to put food out much closer to the hide, bringing them within camera range.

'After a few winters trialling the project, we realised we had the kites onside,' Dad said. 'I remember one day, not long after we'd first got them feeding nearer to us, I was standing with John by the cattle shed and kites were flying overhead. I had this feeling, I don't know what you'd call it – a premonition, or a certainty maybe – and I said to John, "That's part of our future up there."

'Until then we'd only been taking groups out on the weekend. It really was the most part-time of part-time jobs. With the kites showing less fear of us, and more people enquiring about visiting, it was time to take the project to the next level, get a proper hide and start daily tours.'

In 2003, the new hide was built in a field above the farm, replacing the old one which has since rotted into the ground. Mike joined the team and the daily tours duly began. Disaster very nearly struck on the very first day when two buzzards diced with death.

'They caught sight of the food but, being a bit wary, decided to perch on a nearby electricity pylon. I suppose they wanted to sit and stake it out for a bit before coming down,' Mum recalled, cringing. 'They must have touched both wires with their wings and created a current because there was an enormous bang and they fell like stones. I was sure they were dead, but both got up and flew away. The next day we got reflective disks put on the electricity wires. They scared the birds, stopping them landing. It was plainer sailing from there and, free from fireworks, we began to really teach people about the kites!'

When the RSPB had begun the reintroductions in 1989, one of the main aims of their action plan was to provide public facilities for kite viewing. If people saw the birds and learned more about them, they would surely have no wish to harm them and no tolerance for those who did.

By 1998, three of the nineteen kites released in Central Scotland in 1996 had died of suspected poisoning. In 1999, a breeding female died of rodenticide poisoning. Kevin Duffy later recovered the skeleton of another bird, most likely her partner, which had almost certainly died of the same cause. That took the death toll to five. (Sadly, unintentional poisoning was a common problem in those days. Many poisons remain in the body of rats after they have died. Kites, unafraid to scavenge around houses and farmyards where rodenticides are most frequently used, are often the first to find the rat and can also be poisoned as a result.) In 2001, two breeding pairs disappeared in mysterious circumstances. They were believed to have been killed.

With the twin problems of deliberate persecution and unintentional rodenticide poisoning so badly hampering the reintroduction, further public engagement was needed.

The idea of kite viewing – protecting the birds not by keeping their existence a secret but the reverse – stemmed from the success of the osprey centre at Loch Garten, near Aviemore. Like so many other raptors, ospreys were persecuted relentlessly in Britain. Unlike the others, ospreys feed almost exclusively on fish. Given the former abundance of river stocks, why would anyone object to a raptor taking their share? However, any kind of loss, be it game bird, livestock or fish, was simply not tolerated in those times. The last breeding ospreys recorded were in 1916. There followed a thirty-eight-year absence until, quite unexpectedly, a pair of Scandinavian ospreys turned up at Loch Garten in 1954. The RSPB set up a well-publicised nest protection scheme, with teams of volunteers camping at the foot of the tree to ward off would-be persecutors and egg collectors. In that first year alone, 14,000 people came to see the birds. So began the concept of putting an endangered species in the public eye.

'We were always looking for a farmer or landowner who might set up a viewing station,' Lorcan O'Toole told me recently. 'I remember the first time I met your folks I mentioned that if the kites stayed here there might be a business opportunity for them down the line. They seemed like the right kind of people, with that kind of entrepreneurial spirit to start a project like that. We didn't hesitate to enter into partnership with them.'

Lorcan is now back in his native Ireland, where he's worked on kite, sea eagle and golden eagle reintroductions. Each species is making slow and steady progress, though reaction to the projects has been mixed and several birds have been killed.

'It's like the early days with your kites,' he said. 'When people around Argaty heard that your parents were feeding the birds, I'm sure some were pretty opposed to it. There were all the old rumours of them taking full-grown pheasants. I used to take people up to your folks' place and show them the kites so they could see how dainty they were and realise they weren't going to be a danger. That engagement factor was crucial.'

We broached the subject of mad cow disease, that terrible back-drop to the start of the kite project.

'In a way it's not too different to what farmers are facing now,' Lorcan said.

'You're right,' I replied. 'It isn't.'

He was referring to the 2016 referendum, when Britain voted to leave the European Union. The topic had dominated conversations ever since.

'Farmers might face real difficulties exporting meat to the Continent,' Lorcan continued. 'Just like they did then.'

Thinking back to the circumstances that had led Mum and Dad to create Argaty Red Kites, I could see the similarities with my current situation. The shadow of BSE had loomed over them then, and a similar one hung over us again now. Lorcan was right, leaving the EU could prove very damaging for British farming. Perhaps the timing of my search for squirrels was fortunate. Farming was facing an uncertain future as a result of Brexit. Meanwhile once common farmland species were in serious decline. Maybe it was the right time to try and diversify the business, time to help our wildlife a little more too. Perhaps history was repeating itself. In the face of one crisis Argaty Red Kites was born. In the face of another, an opportunity was there, for us and for the animals.

4

Unexpected Visitors

So common is the kite with us, that its various modes of flight are considered an almanac for the weather, and its note is the symbol of moral conduct; we observe, when it forks high in the air, it prognosticates good weather; and every boy will tell you that it's not for nothing that the glade whistles; alluding to the note of that bird when it glides through the air, watching its prey.

The Revd Mr James Lapslie, Minister, *Statistical Account of the Parish of Campsie*, 1795

For the prospective customer, the idea of watching squirrels and kites in one location had to be attractive. The idea of dedicating my days to these incredible animals was very appealing, too; the definition of time well spent.

Looking through trail camera images, I was surprised at how dark the first intrepid visitor to the feeder was. The squirrel's back and tail were almost black, its legs and face a rusty brown. Throughout the day it returned frequently. Soon others came. Some were as dark as the first, others so orange you'd swear they'd been dipped in a vat of Irn-Bru. We had squirrels!

The joy that those pictures brought me is hard to put into words. In other respects, however, I was in a mess. In mid-August Mike left and I missed him dreadfully. Wherever I looked I saw the empty spaces where he ought to be, the little things he'd left behind. A discarded name badge. His old tattered camouflage jacket. His picture on the visitor centre wall, above the ironic caption: *Current Staff.* Any tiny thing could set me off. My best defence was to think of the squirrels and the possibilities they offered.

A surprise awaited me when I entered the woods one morning in early September. A strange creature had appeared the night before. Entering night vision mode, the camera had produced a picture so brilliantly white that it was hard to make out much. The animal was as big as a cat – at first I thought it *was* a cat, but the ears were too rounded, the body too lithe. I looked closer. It was a pine marten.

Pine martens are members of the mustelid family, the order of long-bodied, short-legged mammals that includes badgers, stoats, weasels and otters. I don't recall ever hearing mention of them during my childhood, so rare were they. By the early years of the twentieth century they had been hunted to near extinction in Scotland by fur trappers and gamekeepers, and have only begun a revival in recent times.

Their history is not unlike that of the kites'. Both were demonised for their predatory ways. The *Coventry Herald* of 1909 aptly illustrated this. Writing of the martens' decline, the journalist acknowledged that the animal 'dwelt in the woods, where it was hunted with dogs and driven from tree to tree and from rock to rock until it was finally despatched, ignominiously, by men armed with iron poles'. He went on to describe them as 'stealthy, long bodied' creatures, 'like an enormous stoat or polecat' that roamed 'on bloodthirsty rounds'. A few years later *The Globe* reported, 'In the three years ending 1840 no fewer than 246 specimens were destroyed in Glengarry', but again the pine marten is seen as a despicable murderer deserving of punishment.

'Their partiality to mutton was the cause of their undoing,' added the writer. 'Shepherds accused them of killing great numbers of sheep, and there is no doubt that the charge was well founded.'

Note the adjectives. Stealthy, bloodthirsty, enormous. Consider for a moment this question: how could a creature the size of a cat kill a sheep? More importantly, given that said creature is an omnivore, capable of eating berries and nuts as well as meat, *why* would it go to the effort of killing a sheep?

Britons were inclined to state such mistruths with utter conviction and to assign human traits to animals as a means of underlining their point. The kites are also interesting in this regard. We not only imbued them with human traits, but also ourselves with the traits of the kite. The term 'greedy gled', for example, was applied to man and bird alike.

I found some interesting references in the *Scottish Gallovidian Encyclopaedia*. Among them was the 'Auld Poem of the Rent Day', dated 1876.

> The tod's* a beast no easy fed,
> Lykewise the burd they ca' the gled,
> The wasp, the speddart*, and the ged*
> Are greedy curses;
> And factor Jock is damn'd ill-bred
> Wi' our light purses.
>
> *tod: fox; speddart: spider; ged: pike

Also in the *Gallovidian Encyclopaedia* is the term 'Gled's Whistle', a reference to the 'kind of wild whistling scream' kites make when they fall in with prey. 'We apply this, metaphorically,' the encyclopaedia explains, 'to the ways of men, in the phrase, "It's no for nought the gled-whussles": meaning, it is not for nothing that greedy men whistle; it is the good fee that makes the lawyer whistle.'

This trait of ours, to humanise (in the most dehumanising terms) the animal kingdom, continues to this day. Wasps, spiders, foxes, badgers, crows, magpies, buzzards and eagles are all on the receiving end. We are compelled, it seems, to emphasise the wicked, destructive nature of these animals.

Where demonisation led other animals to extinction, the pine marten survived. Just. Newspapers and journals record isolated sightings throughout the nineteenth and early twentieth century. Mostly the marten was spotted and killed, but the fact that they are nocturnal, can range for miles and feed on just about anything saved them from total obliteration. Now protected, they are spreading back across the countryside, back into areas like ours, and a marten craze is growing amongst photographers and wild-life enthusiasts alike.

Increasingly, while I'm on duty in the hide, photographers sidle up and in a furtive whisper ask: 'Got any martens here?' For the record, I have no idea why these animals reduce grown adults to conspiratorial mutterings. There's an allure to them, I suppose, a mystique. They seem to us a secretive animal. Perhaps people like to feel they're in on that secret . . . I really don't know.

Whenever someone enquired, my answer was always 'Yes' – though I knew they were here only because Mike had mentioned them. I had no idea where they could be found. Now I had an answer. In this wood, we had squirrels and pine martens. This knowledge left me giddy with excitement. I had a new mission. I would start tracking them both.

*

I knew I would have to learn more about the martens if I wanted to set up a feeding station for them (and I did want that – very badly). How big was their territory? Did they use set routes every night? Which foods did they like? I had a lot of questions and very few answers.

One problem with the conifer wood was that it was so dark and inaccessible. Although I'd never actually visited a red squirrel or pine marten hide, I knew that none could be located in such insalubrious surroundings. I had to find a bright, open space, somewhere easy to reach. Across the road from the conifer plantation is Drummond's Brae, a broadleaved wood that was my favourite place to play as a boy. The grass outside was thick with nettles and bracken, the terrain within steep and muddy. No adult would ever willingly set foot in there. That made it perfect for a den. At the top of the wood stood a twisted tree. Its branches curled outwards, overhanging the field beyond. I would often scramble up the crumbling slope, crawl through a boy-sized hole in the rhododendrons and climb that tree. Many hours were passed sitting in my little nest, watching wood pigeons land in the branches above me. The pigeons, those plump little ministers in their pink cardigans, seldom noticed me there. We may look for birds in trees. Birds don't look for us up there.

When did I last set foot in that wood? Years ago – decades perhaps. Long enough to forget, in any case. I was clambering up the bank, surprised at how difficult it was to keep my footing, and had almost reached the top when the soil gave way beneath me. My arms flailed, searching blindly for a branch, a tree trunk, anything that might prevent a tumble back down the hill. And they found something very foreign to that environment: a tattered blue rope hanging from a tree. Pulling myself to steady ground, I remembered that my dad – after some cajoling – had waded through nettles, heaved his way up that bank and climbed the tree to tie up my very first rope swing. At five feet, my mum was too small for the task, so she was left on the edge of the woods, hacking back the nettles to make his exit a little less painful. I was amazed that the rope had lasted so long, and even more amazed that I had forgotten it. That swing was my pride and joy. Each and every one of my friends was dragged out here and made to demonstrate their courage by swinging out over the precipice.

Adjacent to my old lair is a disused paddock which backs onto our car park. While the woods themselves are still inaccessible for your average athletically challenged adult – and I reluctantly included myself in that group now – if squirrels or martens would first feed here, perhaps they could later be enticed into the paddock. That would be a fine place for a hide. I fixed a feeder box and camera to trees within my old lair and withdrew.

On my next visit the camera had some delights to show me. Not one but two pine martens had visited. They quickly discovered a liking for nuts and became regular visitors. From the images it was evident that one must be a female and the other a male. The latter had a much thicker skull and larger body. If she was all slinking, feline grace, he was a prowling brute. In the dead of night they would stalk slowly down the rhododendron branches, flip open the box and dip their heads inside. They'd rotate as they ate, constantly checking their surroundings, guards on a constant state of high alert. I'd heard it said that martens could be skittish and prone to flight. Here, in these grainy nocturnal videos, I saw the proof of that.

It seemed possible that one day we might set up a squirrel and a marten hide, although I was realistic enough to know that neither would happen immediately. I would have to draw them slowly out of the woods first. For the time being, I needed other things to occupy me.

In the early years of the kite project we ran wildlife events at the farm. My parents fondly remembered those bat walks and dragonfly days. Over time these had fallen by the wayside. Nobody knew why. They were just other victims of our project's stagnation. It was time to restart them. We had a facility here, an untapped wildlife resource, but we needed experts on board to reveal its potential. As luck would have it the first two charities I contacted, the Bat Conservation Trust and Saving Scotland's Red Squirrels, replied instantly. So far, so easy, I thought. In September, mere weeks after those first emails, we hosted both a bat night and a red squirrel walk. We were back in action.

At the latter event I announced my squirrel hide ambitions to everyone. In hindsight this was a premature step, but I was high on the increasing number of squirrels visiting the feeders and consumed by the idea of the hide. Another reason for telling the world was that I worried that if I didn't I would talk myself out of pursuing the plan. Frequently I'll start something, grow obsessive over it, reach a point where I can see it coming to fruition, then convince myself it's a bad idea and pull the plug on it. I don't know why it happens. My confidence can be brimming and then, for no reason at all, it disappears. People often tell me I'm afraid of change. Fortunately I'm more afraid of looking foolish and, were I to tell everyone a plan then fail to follow through, foolish I would certainly look.

Soon after the events Dean Bricknell took me on a walk through the woods, teaching me about marten tracking. Dean is a professional photographer who has been a friend of the project for many years. We clambered up the muddy banks into my old den, then wrestled through the rhododendrons, all the time looking and listening.

In the middle of a clearing, he stopped sharply.

'See this?' he said, pointing to the ground.

Three shrivelled orange balls lay in a little pool there.

'Rowan berries,' he said. 'The kits love them, but they eat so many that they go through undigested.'

Close by, he found a line in the dirt. It led through a gap in the rhododendrons and out into a clearer part of the wood.

'Bet that's their trail,' Dean said.

I nodded knowingly, privately wondering how he'd spotted either of these clues. I would have trampled right over both. That was quite a sobering moment. Not for the first time the limitations of my plans began weighing on me. Seeing Dean at work, I realised how much I had to learn.

By that stage it was too late to back out. After a period of negotiation we had contracted out the farm to a charming couple who lived nearby. For six years – or longer, if both parties wished to continue beyond that point – they would look after the sheep and cattle, leaving me to follow this new path. The move had allowed Dad to step back and his relief was evident. I'd never heard him complain and because of that had just assumed he was happy. When at last he was freed from seven-day weeks he admitted that he hadn't been sleeping for a long time due to the strain of captaining the ship more or less alone. Although he never said as much I guessed that he hadn't wished to pressure me, so had never asked when or whether I was going to step up and take over the business. I felt pretty lousy when I thought about that.

Now relative strangers were running the farm. It was a surreal experience. The changes had been highly unsettling for Mum. She hates uncertainty, and the past few months had given her plenty of that. I felt it, too. I've inherited a lot from her – my interests in art, music and literature, the beautiful things in life. She didn't come from a farming background and never really fitted comfortably into this world. I don't suppose either of us did. But for all my years, and the past thirty-five of hers, this life had been all that we had known.

Nothing else for it. Having been the instigator of these changes, the cause of all this disruption, I had to press on with the plan, no matter how imperfect it seemed.

5

Changing Days

This bird, from its great length of wings and tail, is capable of supporting itself in the air with very little motion, and for a great continuance, but is slow in flight; its depredations are therefore confined to such animals as are found on the ground, young rabbits, hares, and game of all kinds, poultry, and young birds incapable of flying. It will also destroy young lambs.

Col. George Montagu, *Ornithological Dictionary of British Birds*, 1831

Autumn came, the trees began to undress. The colours – orange and yellow leaf against green field and blue sky – were spectacular. Kites swooped past the hide, russet bodies glowing against the pastel background. They sent the photographers into a frenzy of clicking.

For the birds, as for everything else, this was a time of transition. The spring, when kites disperse to their nesting sites (where they'll stay through the summer), seemed a distant memory. Winter was just around the corner, you could feel it in the air. The days were drawing in, the migrants – geese, redwing and fieldfare – were back; the swallows had flown south.

That ever-shifting season is always a wonderful time in the kite watcher's year. In those short days, when temperatures drop and food grows harder to find, kites that nested at higher altitudes return to lower ground, joining those birds that nested here in the spring. They form large scavenging groups and circle the country-side, looking for a meal together. Kites are sociable; they feel safe in numbers and don't mind sharing food. These are some of their greatest charms.

When I put the meat out on those early autumn days, one would call to the others, giving them the signal, bidding them to follow. Their call is the sound of a full-time whistle, of a passing ambulance siren. It's a language foreign to most ears, a once-common dialect that disappeared for a time many years ago when the last kite was killed in Scotland. But to me the gled's whistle is the sound of home, and in this strange world that I had created for myself I clung to it, to its familiarity. It reminded me that, no matter how much had changed, some things remained constant and true: this was indeed still my home.

Alerted by their comrades' calls more kites emerged, floating up from the treetops until they caught the breeze, then sweeping forward in a wave. They wheeled above the food, pirouetting in ever-tightening arcs, as water circles a drain. Finding their angle, they swooped. Taloned feet shot out at the final second, the meat was snatched and the kites rose, bloody fistfuls dangling beneath them, paraded like the spoils of victory.

It had always been an awe-inspiring onslaught, intense, like nothing else I'd seen in nature. I'd spent so much of my life watching it and still it astounded me.

In late September the weather turned harsher and more kites fled the hills to join the Argaty flying squadron. Many youngsters were among them. They'd flown the nest just two months earlier, and by the end of the month many of them would leave here. They had no partner, no nesting site. Already their parents had left them to their own devices. There was little to tie them to this area. Most British kites breed for the first time at two years old (although some are precocious and mate a year earlier; some are late starters who take time to find a partner). In the years before they reach maturity, pair up and settle down, youngsters live a nomadic lifestyle, moving around the country looking for other kites and bigger, better supplies of food. Once they hit breeding age, many will return. If they have successfully fledged a nest here, then they will know that the area is safe. Safe to return to. Safe to raise their young. As someone who came back to the place he was raised, this loyalty to the land has always appealed to me. It beguiles me, too. How, after two years on the road, do they find their way back? Is it something coded into them? Do they navigate using geographical landmarks? Does the position of the sun guide them? Or do they possess even greater powers? Recent studies on robins and zebra finches have revealed a special protein in the birds' eyes which allows them to visualise Earth's magnetic field and, possibly, determine their direction of travel in relation to it. Are kites also capable of this?

Nature is full of mysteries, and birds have added a great deal to the lengthy list of things we don't understand. I like that. It's comforting to know that there are things we don't know, puzzles we can't solve.

Some of the young kites we saw in those autumn days would never return here. Some would settle in other areas, some would die. The first winter is hard for them and there are still places where they aren't tolerated. If one thing didn't get them, the other very well might. The visitors to the hide in those weeks were lucky. They may have been the last people to see several of the young here.

The fledglings were beautiful, and close inspection revealed differences between them and the adults. Where the mature kites had striking pale yellow eyes, the youngsters' were darker and deeper, an indefinable blend of blue, grey and brown. When banking against the autumn sun, the white flecks on the juveniles' sandy breast feathers cast a dazzling reflection. (They would wear these colours until the first moult, then assume their adult plumage, dark streaks licking like flames at the centre of chestnut brown feathers.) Another clear difference was how pristine the young birds were compared to their older compatriots. Deep into their annual moult, which they usually delay until their chicks have fledged, the adults looked wretched. Old, worn feathers had dropped like Christmas tree needles, leaving big gaps all over. One year I remember seeing a kite that had moulted out all of its tail feathers. The tail is the kite's rudder. How that bird managed to direct itself was beyond comprehension. A kite minus its trademark forked tail remains one of the strangest things I've ever set eyes upon. For all that their wings and tails were tatty, the adults made diving for food look easy. They swooped at incredible speed, grabbed and went. Even with forty kites circling, you could spot a youngster instantly. They were the ones floating lower and lower, trying to tuck in their wings and dive, but unable to do so. They were the ones that swooped from one metre above the ground and still managed to miss by two. They were the ones that flopped down to the grass, ate so much that they struggled to take off and ended up teetering away, unbalanced and unstable, wobbling fatly into the woods. They never failed to amuse.

Some young kites had real struggles. They kept swooping, kept missing. One day I watched one descend at a horribly ill-judged angle. As it neared the ground it collided with another bird. They squeaked insults at each other in pathetic mousey voices, then flew off in opposite directions. After more failed attempts the youngster gave up, landed and at last grabbed some lunch. Eschewing the chaos of the feeding area, where so many kites were snatching meat, an older bird waited in the wings, watching the scene unfold.

It made a beeline for the juvenile as it flew off with its meal. Taken unawares the young bird panicked and dropped its catch. Soaring beneath, the wily veteran plucked its winnings from the sky just before they hit the grass. Another autumn. Another typical scene.

I'd miss the youngsters when they went. Even after all these years, after so many generations have fledged nests here and then gone, that melancholy feeling still hits me. I mourn their absence. If that sounds strange, all I can say is that some things never change, no matter how much time passes; the emotions that these birds inspire, the wonder, the joy and, yes, the sadness, are evergreen. Besides, perhaps there is something selfish in that sorrow. Mike used to say that when you live and work in the countryside, you mark the passing of time not in the changing of the year, but the changing of the seasons. As another generation of kites leaves this area, I age by another autumn. Often I wonder if it's really the passing of my own life that I mourn.

Amid the kites diving each day, I caught sight of some that still had plastic tags on their wings. For the first twenty years of the reintroduction, every chick that could be found was tagged so that they could be identified from a distance. A small length of dental wire was used to pierce the wing, midway between the shoulder and carpal joint. Then the tagger attached the tag to the wire and tied knots at either end to keep both tag and wire on the wing. (In the early years strimmer cord was used, but this turned out to be biodegradable under ultraviolet light and many of the tags fell off.)

Those plastic tags were, in effect, red kite passports. Each British reintroduction site had its own left-wing tag colour – a red tag indicated that the chick hatched in our area. The right-wing tag colour showed the year of hatching. They'd be pink one year, blue the next, moving on through the rainbow as time progressed. If you looked very closely you'd also notice that the tags had numbers, letters or symbols on them. These codes, unique to each bird, were used to tell individuals apart. Tags were fitted when the chicks were still in the nest, aged roughly four weeks old. (Once they'd fledged there would be little chance of catching them; for all that kites are accepting of people, there's only so close they'd let you come.)

Much of what we have learned through wing-tag sightings contradicts what people thought they knew about the birds. For example, while kites in Scandinavia migrate to Spain to avoid excessively cold winters, it was previously believed that ours would remain in the UK year-round. Thermal currents are far stronger over land than water and kites have always struggled to fly without a breeze to propel them. Surely, we told ourselves, they wouldn't put themselves to the trouble of crossing the sea when British winters tend to be mild. The birds challenged our half-baked assumptions, turning up in all sorts of weird and wonderful places, from the Outer Hebrides and Ireland to far-flung and exotic destinations such as Normandy, Asturias in Spain, and Porto in Portugal. For a short spell, the kite that flew to Portugal held the record. Its 1,054-mile flight in December 2003 was the longest known migration ever made by a UK kite. (A kite from the Inverness-shire population later trumped this by flying all the way to the Azores.) Thanks to the tags, we learned of those birds that had emigrated. People in those far-off places saw these little strips of plastic and word filtered back, first to the RSPB and the Scottish Raptor Study Group, and ultimately to us. Tagging is a remarkably simple idea, but so effective.

The chicks of 2016 were the last to be fitted with the wing-tags. By that point there was no sense in continuing. There were so many nests and the breeding territories were so widely dispersed that in the final years only a proportion of youngsters were tagged. Compare that to the early days, where every nest was known, and every chick was tagged, and you get an idea of how successful the reintroduction has been.

The decision to stop tagging made sense in another way, too. Bird recording is largely carried out by Raptor Group volunteers outside of work hours. Fitting tags used up precious time that could have been dedicated to finding and recording other species, many of which were faring worse than the kites.

I understand this, and know that ceasing to tag was a sign of progress, but I miss those early days. I miss receiving the reports of sightings in those strange, unexpected places, miss discovering that birds tagged in other areas had appeared at Argaty. We do still fit leg rings to as many nestlings as we can find. (These metal rings, shaped like a letter c, are placed onto the leg, and special pliers are used to close the open side, moulding it into a perfect circle. Each ring is engraved with an alpha-numeric code.) The trouble with the rings is that they are so small that the codes can't be read from a distance. If a kite is recovered, injured or dead, and the ring number is checked against the central database we learn a certain amount about the bird (for example, how long it lived and how far it travelled from its place of origin), but we'll never again know what happened to it between the start and end. This gives me a very real sense of empty nest syndrome. We spend our spring and summer finding and monitoring nests, we leg-ring the birds, watch them leave the nest and come for the food. We are with them all the way. Then suddenly our link to them is broken. They disperse and we don't know what happens to them next. I suppose all advances come with some cost. Each year I tell myself just to enjoy the birds while they're here.

6

Impasse

There was shot on Saturday last, by Mr Wood, game-
keeper, Eden, one of those birds so rare with us nowadays
– hawks called the kite . . . The one in question, a male, is
a splendid specimen.

'Local Intelligence Banffshire', *Banffshire Journal
and General Advertiser*, 5 May 1857

As the days drew in, the last of the leaves fell and I continued my
woodland experiments. Mum joined me sometimes and we pored
over the camera pictures together, both of us delighted to have
found this window onto a secret world.

We talked a lot on these walks, discussed our new lives, discussed
our old. I was glad of her company, glad of the conversation.

'Did you ever hear about the time we became home to a population of gannets?' she asked once as we made our way through that
shadowy spruce maze.

'Gannets!' I said, turning to stare at her with incredulity. 'This
far inland?'

'I know,' she agreed, laughing. 'I could hardly believe it either.'

She paused to push aside a razor sharp eye-level branch, then
continued. 'A group of old ladies used to come to see the kites each
year. They were a hilarious bunch, very well-to-do and utterly full
of mischief. The young woman who ran the trips had a bit of a

job keeping them in order. On one of their visits the kites were taking forever to come for the food. The ladies got bored and, much to the organiser's horror, started passing around a hipflask! I left them to it.

'Later I met them coming down from the hide, happily announcing that they had just seen a gannet. I thought to myself, "Whatever's in that flask must be good," but they were insistent, so I went looking and, of course, found a gannet lying in the grass a few metres from the hide. The poor thing really did look like a fish out of water and was making no moves to get away, so I ran to the house and called the SSPCA. Apparently ours wasn't the only bird who'd ended up in such an unusual spot. There'd been some terrible winds the day before and hundreds had blown inshore. Gannets can only take off from cliffs – something we're obviously a bit lacking in here! Given our central location, people began bringing gannets to us for the SSPCA to collect. I came down to the car park one day and a man was sitting in his car, patiently waiting, with one on his knee! I don't know if I'd have felt that comfortable in such a confined space with that big beak. The SSPCA came out and collected the whole lot, took them to the seaside and released them on the clifftops.'

I shook my head and smiled. My parents amused me sometimes. So many amazing and unlikely things had happened since the project began and yet they could go years without mentioning any of them.

On milder days I began taking Rowan out on my walks around the woods. Although only three years old, she was a great little helper. We'd look at the stealth cameras together and, if I lifted her, she was able to top up the feeders. She was a constant delight, gathering skulls, feathers and other treasures and taking them home to add to her collection. Her fascination in anything new was a beautiful thing to see.

To this day Rowan and I share something special, something

born in those days when we wandered the woods together. We have a mutual curiosity about the natural world, a wish to uncover its mysteries. We can pass hours searching for insects beneath rocks, watching birds in the garden, fishing little creatures out of ponds, inspecting them and returning them. Ours is a friendship based on a shared passion.

One of the great failings of education in Scotland, as with much of the developed world, is that we don't make the time to properly teach children about nature. We wedge it into the curriculum in a tiny gap between biology and geography, spend a few spare hours learning from textbooks and consider the job done. In my youth I was taught so little of the natural world. It angers me now, at this later stage in life, because I have to learn everything for myself. What knowledge I have feels hard-won and there are glaring gaps in my understanding.

I hope Rowan never feels this way, never feels lacking, inferior, less-than-whole. Ever since those first walks in the woods I've tried to share everything I learn with her. I take her with me whenever I can so she can truly experience the world around her. Nature is fragile and a wrong move by us can cause such damage. On an intuitive level Rowan knows this, and that means a great deal to me. One day – if she wants it, wants to make a life here – she'll be responsible for this place, and all of the life within it. I hope she'll feel that she has enough knowledge to make a go of it.

*

Come October, with the weather worsening, the days darkening, I came to a conclusion. I'd had it with battling into the darkest depths of the spruce wood looking for squirrels; with its treacherous muddy banks, the marten site across the road was little better. The time had come to lure these creatures to more civilised places. I'd scotched the idea of building a single hide for viewing both squirrels and martens: the former resolutely refused to cross the road to the latter's feeding area. Two hides would have to suffice. If I could entice the squirrels to a clearing in the conifer wood I'd build one there, while the paddock beside my old lair still looked like a prime site for the martens.

The trouble was that neither animal would move. I shifted the feeders a little every day, but it was to no avail. Anywhere other than the murkiest, most inaccessible place and they simply wouldn't budge. Each day I checked the cameras. No pictures. The peanuts in the boxes went mouldy and had to be thrown out. Unable to find a way forward I moved the martens' box back to its original position. They returned instantly, visited regularly for a week and then, for no reason I could see, disappeared. I've seen no sign of them since.

A terrible depression set in. I had no idea what to do. Mike came up to stay, his first visit since leaving, and though it was great to see him it reminded me of how much I missed him. The excitement of the squirrels, martens and of the events we'd hosted had been wonderful distractions, but in truth I was still badly bruised, just as badly as when he'd left. I said nothing to him of any of this, nothing of my failures at work either. I'd wanted to have something to show for my efforts, some proof that the faith he'd once had in me had not entirely been misplaced. And anything else I might have told him – that he was my brother, that I loved him and missed him – I didn't, because it all just hurt too much. His visit came and went and I carried those wretched feelings onward.

Day after day I returned to the woods. The result was always the same. No food taken, no pictures on the cameras. I built more feeders, put them up in other places. None of them was used.

'What you up to today?' the other men on the farm would ask, pausing between jobs that I should have been doing.

I'd tell them. They'd fail to think of anything to say. They'd walk away.

'I'm not being funny, but is this actually going to make you any money?' one asked one day.

'Yeah,' I said, without conviction.

I was, as I'd most feared, starting to look foolish. And it hurt. It truly hurt. At one's lowest points life can start to feel like a long and tragic metaphor. A little boy, lost in a dark wood, desperately chasing after things that had run away from him. That was my tale. You don't have to dig too deep to find the meaning in that.

November came. Great skeins of geese arrived, scrawling shapes and letters on the sky in a wavering hand as their formations constantly shifted. In a bid for progress I spent several days emailing conservation charities, suggesting we work together, form partnerships, co-host exciting events. Those that replied said they'd be in touch the following spring. None wanted to do anything in that cold, wet November when I needed it most.

Dean Bricknell had left a pop-up photography tent in the paddock we'd hoped to lure the martens into. One evening snow began to fall heavily. It gathered on the roof, the pile grew higher, the tent caved in. My mind saw a thousand more metaphors in that.

If I thought I was bearing all of this with silent stoicism, I was wrong.

'It strikes me that your life these days is rather lonely,' my mum said one day.

My eyes began to fill. Muttering some reassurances, I walked away in a hurry.

From the tall pines outside my parents' house, the clever starlings wolf-whistled as I passed. Their mocking calls mimicked the kites that circled above. Then the tune changed and, ringing out from those treetops, echoing across the yard, came my father's dog whistle. A kite call and a farmer's whistle. A cruel reflection of a life still available, just, and another gone, given away, lost.

Breaking Through

7

New Year

Advancing civilization has done its work with the kites, as with all other destructive animals, and driven it far away from human habitations. Man chooses to be the only destructive animal within his domain and, eagle-like, permits no inferior to poach on his territory.

Elgin Courant & Morayshire Advertiser,
31 August 1860

The winter of 2017–18 was the harshest I've ever known. Snow followed rain, rain followed snow. Pining, skeletal deer shuffled around the frozen estate, searching in vain for a bite to eat. Several died. Some I chopped up for the kites, others I left for the foxes and crows. Buzzards succumbed to the elements, too. I'd find them in the fields and woods, their bodies so thin that their breast bones jutted out like Adam's apples.

The field around the kite hide is filled with rowans, hazels and thorny bushes which provide food and shelter for small birds. With help from the local RSPB members' group, my parents had planted them back in 2003. That was the year when they abandoned their old homemade hide and had a new one built closer to the farm. Usually these trees and shrubs have enough berries to last long into winter, but not in that everlasting cold. Each tree was picked clean, a carcass stripped of meat. I set up several bird feeders, and the starving tree sparrows, chaffinches and tits descended with delight.

As pond upon pond froze over, herons began visiting the kite feeding area. Heads bobbing to some imaginary beat they dad-danced across the grass, Bobby Charlton comb-overs blowing in the wind. Their presence drove the kites crazy. They bombarded the meat, afraid that the alien invaders would devour it all.

Thirty to forty kites normally congregate in the Argaty winter roost, but by the early weeks of 2018 the number had grown much larger and we saw more each day as the big chill continued.

Watching the birds in winter, I found myself wondering why there was no collective or communal noun for a group of kites. We have charms of goldfinch, parliaments of rooks, gaggles or skeins of geese, murders of crows. Kites roost together, fly together, find food together. They are the very definition of a communal bird, and yet nobody has seen fit to give their group a name. True, some refer to soars or kettles of kites – the term apparently referring to the birds circling together like liquid boiling in a cauldron – but these are applied to every other social raptor too. Don't kites deserve a title of their own? It's not as though people are unfamiliar with them. Quite the opposite is true. Songs and poems have been written about kites, the birds lent their name to the popular toy, and in Scotland the towns of Gladgate, Gladsmuir (kites' moor) and Gledstanes (kite stones, from whence William Gladstone's family originated) all owe their name to the birds that roosted nearby. So why no collective noun? For posterity, here are my humble submissions: a whistle of kites, a summons, a tiding, a tryst, a garland, a torque, a spiral, an eddy . . . I could go on. The point is that our ancestors weren't short of choices.

Living in a pack is a sensible survival strategy. Winter is hard. If the ground is frozen, digging up insects is impossible; if snow is heavy, any carrion soon gets buried. Many pairs of eyes are better than one and for kites, who have slow metabolisms, sharing a meal with their mates isn't a problem. Those birds that nest nearby tend to stay on Argaty over winter. They know there's food here, know it makes sense to remain. They are usually the first to appear when we throw the meat out. Non-resident birds, recently returned to the communal roost, notice the locals heading purposefully towards the hide each afternoon and realise that they are queuing for a reason. Within minutes, the whole team is diving for the food. Working in this way, they help one another through the worst of the weather.

One day in early January, with the roads so thick with snow that no visitors dared drive them, Mum and I watched sixty kites skate across the frozen sky and descend. I'd never seen so many at one time. In the midst of the action, a lonely buzzard stood hunched over the meat, fighting gamely for its share. It was a youngster, searching for a territory, desperate for food in that merciless weather. I felt for it, I really did. It wouldn't last long here. A nesting pair of buzzards patrol this spot and they don't tolerate interlopers.

The young bird had been there no more than five minutes when a howl drifted down from the trees on the hill. One of the adults had spotted it.

For two more days it remained in the area. The poor thing must have been famished. Each time the kites swooped it reared up, flapping its wings to shoo them away. Then it ducked its head and gorged on the meat again. For that short period it was in heaven.

On the third day, the territorial male came. He left the trees at mid-height, passing unnoticed through the throngs of kites diving for food. At the last second his talons shot out. The collision was utterly silent and somehow more sickening for that. The youngster fell like it had been shot and lay there motionless. I watched for several minutes, convinced it was dead. At last, as I was walking over to gather it in, it dragged itself up, limped several paces from me then launched into a punch-drunk flight, disappearing from sight, never to return.

Buzzards have endured a rollercoaster ride in recent decades. By the 1960s their numbers had reached terrible lows. First they'd faced starvation after the disease myxomatosis had decimated the population of rabbits, one of their main food sources. Then widely used agricultural insecticides such as DDT hit hard. The poison travelled up the food chain, killing not only the insects but the animals that fed on them too. Ingesting even a small amount had another terrible side effect: it thinned eggshells. When the female settled down to incubate, the shell cracked beneath her weight. The poisons were subsequently banned and buzzard numbers are rebounding. Today, at an estimated 60,000 pairs, they are Britain's most common bird of prey.

In the past few years, however, localised declines have been reported. Some are in areas where kites were reintroduced, and although the number of buzzards in Britain is more than twice the entire global kite population, many blame kites for the problem. Why, they ask, did people bring these birds back? Why are we feeding them? Surely they should be left to fend for themselves. A few go even further, suggesting that kites should be culled to redress the balance. For anyone involved in kite conservation, these questions are pertinent. If we wish to show the importance of kites to our ecosystem and prove that our work is more than some mere glory project, we ought to explore them.

Those involved in the reintroductions would say it was a moral duty to bring kites back. As Roy Dennis, the driving force behind the very first projects, recently told me, 'I had always hankered after red kites after reading of the onslaught on them in Scotland. It was a matter of correcting a wrong. I wanted to see them flying again in their ancestral haunts and am thrilled it has been so successful.'

There are ecological reasons for the reintroductions too. Picture the time when kites were our most abundant bird and circled every inch of our countryside, feasting on carcasses wherever they found them. Think of the many wild animals roaming these lands, saved from exposure to diseases and parasites found in that decaying meat by the kites and their amazing, catholic diets. Remember the people in towns and cities who were spared from a similar fate, thanks to these birds. Imagine all of those kites flying in to roosts around the country each night, gathering in the treetops, defecating and coughing up pellets, carpeting the forest floor in an ultra-fertile mix of bone meal and manure, helping everything to grow. All of this was lost when kites were removed from our landscape.

And what of today? What role might returning kites play in our very changed countryside? Where once they saved wild animals from contact with carrion-linked diseases, kites can now save healthy sheep. When a ewe dies on a hill farm it's a lottery as to whether the farmer will be able to find and recover the carcass. The sharp-eyed kites never miss a free meal, however. What a benefit this could be, both to the remaining livestock and to our food hygiene. We should also consider the other scavengers: corvids and buzzards, badgers and foxes. These are capable hunters, and in our degraded landscape their impact on prey species can be severe. With kites back and competing for carrion, might these rivals be less prolific? Might instances of predation reduce? If the aforementioned anecdotes are anything to go by, in some areas kites are already affecting the balance. For those people used to seeing larger numbers of buzzards any recent decline may seem alarming, but nobody can say how many kites or buzzards there should be in Britain. In most areas, the only avian challenger the solitary, territorial buzzard had in the twentieth century was other buzzards. Although that might have seemed natural, it was anything but.

Nature intended Britain to have scavengers co-existing, some-times competing, but ultimately finding a balance. Beyond any moral considerations, balance is the real reason for bringing kites back. These birds help draw us closer to the healthier equilibrium we used to have.

The second question – 'why feed kites?' – is easier to answer. We feed them because they are scavengers. Because, deliberately or not, people have supplied much of their meat for centuries. Even if all feeding stations were abolished, kites would not simply disperse and live independently of mankind. They are not that kind of bird. Instead they would go to busy roads, railway tracks, wind farms and shooting estates, the places where we regularly kill things. And as a pack animal loyal to their territory, they would go en masse. Short of outlawing cars, trains, wind turbines and field sports, there is little we can do to change that. This being the case, surely we'd prefer to see kites at feeding stations than at those dangerous places.

That said, we must be careful not to overfeed them. These birds are opportunists. The more food you provide, the more kites will come. At Argaty we put out less than one kilogramme a day (an amount intended to replicate what they once found in the wild in the days when farmers left sheep carcasses out for the birds). Our food maintains a small population only, provided that they hunt and scavenge for their own food too. Living alongside them are buzzards, songbirds and so much more. We strike a balance. Overfeeding risks seriously upsetting that. Looked at properly, the issue is not *whether* or not to feed kites, but *how much* to feed them.

The final question – 'ought we to cull kites to redress the imbalances they've created?' – is the one that requires the most in-depth response, for it takes us to the heart of Britain's wild-life crisis. The truth is that kites have not created an imbalance.

Instead they have entered an unbalanced world. While at places like ours – traditional old estates with plenty of trees, untamed areas, hedgerows and ponds – they have slotted back in with minimal trouble, that isn't the case across all of Britain because so much of our countryside is no longer wild. Although we have been altering this landscape (and impacting wildlife) for most of our existence – felling trees, draining wetlands, hunting and killing our way through the centuries – since the Second World War agriculture has intensified to previously unimaginable scales and we have changed the countryside hugely. Now we have so little habitat, so few unkempt places where nature can thrive. This is what creates conflict between species.

A major cause of our problems is the UK population, which has risen from around 49 million people in 1945 to over 66 million today. Much of that growth came in England, where there are nearly twice as many people as there were at the start of the twentieth century. Despite our small land mass, our little island now has the third-highest population in Europe. Eurostat, the EU's statistical office, claim that if the trend continues by 2050 we'll have the highest. The pressure this puts on our countryside may not be immediately apparent. Only 17 per cent of Britons live in rural areas, and more than 70 per cent of our landmass is farm-land. Our population may be growing, but it is not eating into our countryside, at least not in a direct way. There is still plenty of space where nature could, in theory, flourish. Our hungry mouths bite in other ways, though. Because our population is growing we must farm hard, simply to feed people. To make matters worse, as a society we've also become wasteful. Every year 25–30 per cent of UK farm produce ends up uneaten in the bin, and we farm our best land all the harder to cater for this waste, spraying more weeds, applying more artificial fertiliser, flogging it until it has nothing left to give.

For decades, European agricultural subsidies have exacerbated these issues. The aim of these annual payments was noble enough. They encouraged our farmers to produce and allowed us to compete with the New World Order – countries like Australia and New Zealand, who were already operating intensively, on vast scales, and driving down prices. If our farmers had charged according to the cost of production, we'd have been priced out, but with the government making up the shortfall, our produce could also be sold at what supermarkets deemed 'fair prices'. Food should be both plentiful and affordable for all. Subsidies made that possible.

But the system had unintended consequences. First farmers were supported per head of livestock they owned. With farm gate value reducing and production costs increasing they overstocked simply to claim subsidy and, often, to survive. An overabundance of food was produced and much of it was, inevitably, wasted. In attempt to tackle these issues, policy was rewritten and farmers were instead paid for every farmable acre they owned. Suddenly it didn't matter if you actually *had* any livestock; you were paid simply for having a farm. And the bigger the farm, the better, the more money you claimed. For other farmers, these changes were a death knell. They could no longer make anything of their small patch of land and gave up.

The wild animals that had evolved for centuries alongside those people were suddenly abandoned. The best land was bought up, the small farms swallowed up by bigger ones. Men with spades arrived and planted the poor ground with foreign trees. A few hardy birds and mammals clung on; the others dwindled.

Across Britain, livestock numbers have dropped each year, but because farmers have to keep land in farmable condition in order to receive those vital payments, wildlife populations have not risen. They couldn't. The habitat simply wasn't there for them.

Today woodland covers only 13 per cent of the UK's land area. The European average is 37 per cent. We have 500,000 miles of hedgerow now; less than half of what we had before the Second World War. Gone are 97 per cent of the wildflower meadows and 90 per cent of the wetlands we had in the 1930s. According to the 2013 State of Nature report, 56 per cent of Britain's species declined between 1970 and 2013, with 40 per cent of that total showing moderate to strong declines. Of farmland birds, we've seen a 54 per cent decline, with 12 per cent threatened with extinction. Among 218 countries whose biodiversity was measured using the Biological Intactness Index, Britain ranked 189th.

We needn't remain in this mess. Marginal land could easily be taken out of production. We could restore and repair habitats. But for so long the system has encouraged us to sit back and watch the decline. In this brave new world so many of us benefitted. The farmers that could go with the changes were paid; they stayed in business, kept on farming. The consumer got cheap food, and those who could really have afforded to spend more on it now had disposable income to throw at other things. Supermarkets and the agri-industrial complex grew richer and richer. We all won. But in this brave new world we also lost. So many farmers are now slaves to subsidy and the supermarket. Repeated chemical doses have wrecked soils. Consumers have lost touch with the land, lost the ability to produce food and become reliant on agricultural methods that can't be sustained. Only the mega-businesses are winning, and in time even they will lose because the system is heading for collapse.

What all of this means on a practical level is that birds like buzzards may very well be struggling in certain parts of Britain, but look beyond the kites, beyond the obvious answer, and the real reason is revealed. There are fewer formerly common birds like sparrows around, too, and when we drive around on a warm summer night fewer insects crash onto our windscreens. All of this is related. None of it is coincidence. We are all to blame. We've gone to war with the countryside and we've trounced it. And our wild animals? We've crammed them into the few remaining wild areas and forced them to fight it out. Culling our native species is not the solution. It is a tool to prop up this lousy status quo, a means of ignoring the damage we do when we blindly support a system that wastes land and food.

Pity the insects, the small birds and so much of our wildlife. Pity the buzzards, too. As I watched that injured juvenile disappear limply over the hill I certainly pitied it. As landowners we had been lucky. We inherited Argaty rather than having to buy or lease it. Our farm was big enough to qualify for a decent amount of subsidy. Our ground was largely unsuitable for the sort of 'improvements' that swept other parts of the country. We had diversifications which always kept some money rolling in. These factors combined to give us degrees of financial and farming autonomy which others didn't have. We could choose how to work our land, which modernisations to adopt and which to reject. Crucially we could *afford* to maintain wild places, to make a home for some species. But we could only provide a home for some. The rest had to take their chances. I hope that the food we provided was enough to keep that buzzard going, that it found its own territory in the end and survived the winter. It's a hard world we have created for these birds.

8

To Kill a Killing Bird

An hour's good climbing brought us in sight of the
hollow frequented by the stag. We had before made
several pauses to reconnoitre with our glasses, as we
mounted from one steppe to another, during which
Walter brought down a very fine hawk, which Donald
pronounced to be a young gled, its wings measuring at
least five feet from tip to tip.

James Conway, *Forays among Salmon and Deer*, 1861

In January, midway through the long winter, I tried to break a cold
streak of my own. Nothing had happened in my working life for
months. The squirrels were lying low, content to stay in their dreys
and wait out the weather; the martens were long gone; we hadn't
held any talks or events since September. The kites remained as
mesmeric as ever, but our visitor numbers were very low as a result
of the harsh winter. I was going out of my mind with frustration.
Something had to happen.

Finally, much to my relief, I found someone who was willing
to work with us. Andrea Hudspeth, a well-respected ecologist,
agreed to give a presentation on wildlife crime in the UK. She
and her Scottish Raptor Study Group colleague, Logan Steele,
had just won the Political Advocate of the Year award at the
RSPB's annual Nature of Scotland Awards. Earlier in 2017 they'd

helped write a petition calling for a licensing system for grouse moors in Scotland. They subsequently brought it to the Scottish Parliament, presenting first to the Petition Committee and then to the Environment, Climate Change and Land Reform Committee.

In the proposed new system, estates would have to apply for a licence to shoot game birds. If the estate was found guilty, on balance of probability, of committing wildlife crime – examples of which include raptor persecution, nest disturbance, use of illegal traps or snares, hunting mammals with dogs, possession of illegal poisons or baits, etc. – their licence could be revoked. Shooting on that estate would thenceforth be banned for a period of two to three years. These offences would be tried in the civil rather than the criminal courts. The burden of proof required to secure a civil conviction is much lower than for a criminal one, meaning that, in theory, more of these crimes could be successfully prosecuted.

The suggestions came as a response to continual raptor deaths and disappearances in areas managed for a certain type of field sport: driven grouse shooting. On these moors, beaters walk in a line, flushing birds from cover, driving them towards shooters who wait in shelters known as butts. People pay a small fortune to shoot there and expect a lot of grouse to fire at for that money. It's worth noting the difference between driven grouse shooting and walked up shooting. The latter is more like hunting. Usually there are no beaters, and shooters walk the land themselves, often with a dog, and shoot the grouse they flush from cover. Fewer grouse are needed to sustain this.

It had long been suspected that intensive predator control, both legal (killing foxes, crows, stoats and weasels) and illegal (killing raptors), was required to ensure such grouse numbers. The 2017 SNH report Analyses of the Fate of Satellite Tracked Golden Eagles in Scotland provided solid evidence to support that theory. Between 2004 and 2016, 131 young eagles were fitted with trackers which showed their locations at all times. Of these, forty-one birds disappeared without trace. One moment the tag was transmitting data, the next it went completely offline. There was, the report concluded, a strong association between these disappearing eagles and land managed for driven grouse shooting.

The technology had a malfunction rate of just 2 per cent. In other words, the tags were almost completely efficient, but if for some reason they were failing the fitters knew about it from the data supplied. If the bird died of non-suspicious causes, the tag continued to transmit, allowing the body to be found. Dead birds had even been recovered at sea on occasion. The same technology had been used to tag eagles in the United States and in many other countries. There were few issues there. Tags in Scotland were 25 per cent more likely to go offline than the same tags used in these other countries.

Something sinister was happening here. Scavengers weren't eating the corpses – large bird carcasses are rarely scavenged on, and in any case predator control on intensively managed moorland ensures that there aren't many scavengers there anyway. The birds weren't flying into wind turbines – a wind farm employee would surely have reported it. Nobody on the moors found the bodies and reported them either. So where were they going? There was a clear conclusion: the eagles were being shot, their tags destroyed, their bodies disposed of.

The story with hen harriers was horrifically similar. The RSPB believe that there is enough suitable habitat to sustain a popula-tion of around 5,000 harriers in the UK and yet the estimated number is around a fifth of that. In 2013, for the first time in history, no successful nests were recorded in England. By 2017 the number had risen to a paltry three. The Scottish statistics, though better, still gave huge cause for alarm. According to the National Hen Harrier Survey, in 2016 there were 460 breeding pairs in Scotland, 9 per cent fewer than there were in 2010 and 27 per cent fewer than in 2004. (In 2019, researchers from Aberdeen and Cape Town Universities would analyse data from fifty-eight satellite-tagged harriers and conclude that the birds were ten times more likely to disappear or die in areas managed for driven grouse shooting.)

It's worth noting that these are just the tracked birds. Hundreds of raptors aren't fitted with tags. If they are killed and their bodies disposed of, nobody will be any the wiser.

The figures make for a savage indictment of Britain and the practices that are still tolerated in parts of our countryside. Reading them, one thing was clear to me. Our laws weren't working. Politicians had to up the ante if they were to deal with the problem.

Thanks to Andrea, Logan and their colleagues in the Raptor Study Group, it looked as if the Scottish Government might just do that. In November they had set up the Grouse Moor Management Group. Chaired by Professor Alan Werritty, the group would examine the environmental impacts of various moor management practices and consider the option of licensing grouse moors. Having gathered their evidence, they would deliver their recommendations to the government in the summer of 2019.

Keen to learn more about all of this, and aware that our visitors would be too, I duly began advertising Andrea's talk. The following day I logged on to Facebook. To my delight the number of people interested in attending was huge. Many of our regular visitors had signed up. People I didn't know were on the list, too. Some had left comments on our page. Absently I clicked on the first one. My heart sank.

'I want to book fifteen spaces so keepers can come and correct the bile that's being spouted.'

I checked the sender's page. All of the strangers were friends of this man. Our event had been shared several times over and a lengthy discussion had ensued. There was talk of gamekeepers coming up from Yorkshire to show their opposition, of picketing if entry was refused. I felt very, very sick.

My first thought was that we had to cancel. Immediately. My second was that we mustn't be bullied into silence. I bounced between the two for the next few days before eventually caving in. The idea of inviting these people into our home seemed too dangerous. So I was bullied into silence, or silence of a sort. Publicly we cancelled the event, privately we went ahead, setting up a guest list of people we knew and trusted. I'm glad we did. The talk focused on all forms of wildlife crime, including hare coursing, badger baiting and raptor persecution, and I learned a lot from it.

The whole incident left a bad taste in my mouth, however. It was distressing to feel so threatened in our own home, I couldn't even see why these people were so opposed to the licensing proposals. Surely the only ones with anything to fear were the criminals. If I was unclear about that, I was certain of one thing: licensing *was* needed because raptor persecution remained a problem in Scotland.

*

I wasn't always so sure about any of this. Much as it shames me to admit it now, once upon a time the idea that anyone would deliberately kill a bird of prey would have seemed the stuff of conspiracy theories. 'This is the modern world,' the younger me would have argued. 'Surely we've moved beyond all of that.'

This bubble of ignorance was brutally lanced back in 2009, the year I came back to work on the farm and Mike began teaching me most of what I now know about kites. In that formative year I saw first-hand the damage that people do to our birds.

That spring five chicks were taken from nests in Central

Scotland and donated to the RSPB's final Scottish reintroduction in Aberdeenshire. It is a mark of the success of the projects that chicks for the latter reintroductions could be sourced from Britain rather than Europe. At long last we had reached the point where our population was self-sustaining. Mum travelled north to see the young birds released from their aviaries. She took a (very blurry) photograph of them leaving the cages. Those birds flew quickly when they caught a glimpse of freedom. Or so she maintains.

My first thought upon returning to Argaty that August was that the kite project was in great health. The birds were well-established, their range was expanding, and more and more people were visiting the hide. Mike's conservation work was going from strength to strength too. Increasing numbers of barn owls, tawnies and kestrels were using his nest boxes. He'd established an amazing wildflower garden near to the visitor centre, two new wildlife ponds had been made, and we would soon have a grant application approved to plant miles of hedgerows along the perimeters of our fields. Not only would they give our livestock shelter from wind and rain, but they would also link all of the woodland on the southern end of the estate, providing great habitat for birds and small mammals.

All was well on Argaty, and for kites across Britain all seemed well too. According to RSPB reports, in 2008 there were 122 breeding pairs in Scotland alone and an estimated 1,200 pairs in the UK. The population was higher than it had been for two centuries.

That December, four months after my return to the farm, one of the birds we had donated to the Aberdeenshire reintroduction flew south, making a beeline for its place of hatching. Did he know where he was headed? Given that he'd been taken from the nest at a month old and transported to an aviary more than a hundred miles away it seemed unlikely, but with kites you never can tell.

He very nearly made it home. Another day and he'd have reached us and found safety with the other kites here. He didn't get that chance. As he was flying near Braco, some ten miles north of Argaty, somebody blasted him with a shotgun.

A walker saw him crash down into a stream, fished him out before he drowned and brought him to us.

Mike immediately knew that specialist attention was needed. The SSPCA rushed out the moment they received his call. The kite was ambulanced to a vet in Dunfermline, where an X-ray revealed the shocking news: there were no less than five shotgun pellets in his body.

You have to be close to your target to hit it with a shotgun; very close to shoot it so accurately. Whoever had done this must have known the bird was not a pheasant, crow or any of the other birds you were legally allowed to shoot in the UK. As Bob Elliot, then RSPB Scotland's head of investigations, told reporters, this was 'a case of blatant environmental vandalism'. The kite went into emergency surgery. The pellets were removed and his wing (which had been shattered) was splinted and pinned. After that, all anyone could do was wait.

'What'll happen if his injuries don't heal?' I asked Mike. 'Can he go to a falconry centre or something?' How young I was back then. Though a roll of the eyes might have been the more appropriate response, Mike was always patient.

'It's a wild bird, mate,' he said. 'It'll never adjust to captivity. It'd be cruel to even try.'

'So what will—' I began. But I saw the answer on his face.

We waited several weeks. Reporters interviewed Mike, the police appealed for information. Had a rifle been used and the bullet lodged in the kite's body, a ballistics test could have been carried out and the gun that fired the shot identified. Sadly, there's no means of doing that with shotgun pellets. Our only hope was that a whistle-blower would come forward. But nobody did. So that was it. With no further evidence to go on, the police could do little more.

There was some happy news. The kite had recovered from his surgery. The SSPCA transferred him to the care of Stewart Miller, of the charity International Raptor Research and Conservation. A falconer and experienced rehabilitator, he would prepare his patient for a return to the wild.

Recently I telephoned him, keen to discover what he remembered of the kite. So much time had elapsed, and so many birds had passed through his hands since then that, in honesty, I didn't expect him to recall very much. How wrong I was.

'God, yes,' he said instantly. 'I remember him well. He was a character.'

'How so?' I asked.

'Well, for one thing he was the first kite I'd looked after in a long time. Many years earlier I'd been involved in a captive breeding programme down in Wales, so I knew a bit about them, but there was one thing I'd forgotten. It gave me a nasty shock when it came back to me. The SSPCA brought him out and I put him in an aviary. Later that day I went out to check on him and he was lying face-down, totally still. I thought to myself: "Christ, man! You've only had him a day and he's bloody dead!"' He began to chuckle, and after a moment's pause so did I.

'Playing dead?' I suggested.

'Exactly!'

Playing dead – also known in some parts of the world as playing possum, though tonic immobility is its Sunday name – is classic kite behaviour. Every time I visit a nest at ringing time it's the same. Rather than struggle, the youngsters lie stock still, tongues lolling out, utterly unresponsive to external stimuli. A great many animals employ this defence mechanism when they are afraid of being caught or trapped.

'Anyway, I saw him blinking and realised he was actually alive,' Stewart said, laughing again. 'He was always catching me out like that. You'll have seen the hoods falconers put on birds to calm them down? Well, it certainly calmed this boy down. Every time I put it on him he'd fall asleep. I've never known a bird sleep as much as him. I kept thinking he'd died on me, but he'd just nodded off!'

'Was that a result of his injuries?' I asked.

'No. Not at all. People don't realise it, but every bird is different. They all have their own wee quirks. Some are bold, some timid, you never get two the same. This kite's quirk was that he was a sleeper. I was always having to wake him up.'

In the shotgun blast, most of the bird's tail feathers had been blown off. Without his tail, he'd be capable of neither steering nor balancing. This was why he had been given to Stewart. The SSPCA could have kept him for a year, waiting until new feathers grew in, then released him back into the wild, but falconry techniques offered a quicker solution.

'We'd been working with the SSPCA for around five years by then,' Stewart said. 'It took a while to build up trust. Some people have reservations about falconry and the idea of keeping birds captive, but we were able to demonstrate how quickly we could get these injured birds back out there if they embraced this knowledge. With this kite we used an ancient technique called implanting, though most refer to it as imping. Using cocktail sticks, superglue and any feathers we could lay our hands on – some had been moulted out by other kites, some by buzzards, a couple came from Harris hawks – we rebuilt the tail. We put the sticks into the hollow quill stumps where the old feathers had been, snipped the ends off the new feathers, put them on the other end of the stick, then glued the whole lot into place. And just like that he had a new tail!'

Over a period of months, Stewart began teaching the kite to fly to his glove for food.

'Wasn't there a risk you'd tame him by handling him?' I asked.

'Really, that only happens if you release the bird immediately after glove feeding. It'd be like turning a pet loose. When we're satisfied that a bird can fly down and grab food accurately, we return them to the SSPCA. They keep them in isolation for a period, feeding them through small hatches at the back of the aviary. The bird soon forgets about people and the bond is broken.'

On 3 June 2010, six months after the shooting, a large crowd of conservationists and journalists came to Argaty. They gathered outside the hide, watching and waiting. Stewart stood before them, his head dipped into his open jacket. Inside it, sheltered from the wind, was the kite.

'Stewart was hunched down, whispering to the bird, soothing it,' recalls my mum, who tells this story better than anyone. 'The kite had a hood on to keep it calm. I remember feeling quite emotional, watching the two of them. He had spent all of those months working with the bird, bringing it back from the brink. He'd built up this incredible relationship with this wild animal and the moment the hood came off they'd part for good. If that was your job, I'm sure you'd get used to it, but it seemed so sad somehow.'

The story had a happy ending. The following day, when the kites gathered and swooped for the food, Stewart's patient was among them. We recognised him from his wing-tags, which the former red kite officer Kevin Duffy had fitted just before his release, and from his mongrel's tail with its mix of different bird feathers. Down he came, grabbing the meat and soaring off as if nothing bad had ever happened to him. He stayed a few days longer, then continued on his travels. He was the very definition of a free spirit.

Even though a decade has passed since that incident, I still think of that kite constantly. When he was shot I said goodbye to something. What was it exactly? My naivety, I think. For so long I'd held the blind belief that people were fundamentally good, that nobody would hurt something as beautiful as a kite. That innocence is gone. I'll never get it back.

9

Breakthrough

We refer to the indiscriminate slaughter of those
beautiful birds of prey, which, in the language of the
game-keeper, are classed under the name of vermin. To
term an eagle or a falcon vermin, is a stretch of vulgar
ignorance which could only originate in the thick skull
of some low-minded keeper, whose sole ambition is the
pot house boast, that his master has bagged more birds
than the master of some one else ...

Hugh Miller, *The Destruction of Birds of Prey*,
printed in the *Dunfermline Press*, 10 September 1861,
five years after Miller's death

When we talk of once-extinct species now thriving again in
Scotland, we always mention the ospreys and sea eagles first.
Thousands of people come to Scotland to watch these birds every
year. Sometimes I feel as though kites are the overlooked sibling,
the footnote to the tale, and I wonder, why is that?

There's a magic to the osprey story, no doubt about that. This
bird returned of its own volition, with no human intervention,
refusing to be denied its place in our landscape. It was, if you
believe in such things, as if it was meant to be.

Sea eagles have a majesty that no other British bird can match. I
saw one at Argaty back in 2012, a youngster released the previous

year as part of the RSPB's east coast reintroduction. He soared over the feeding area and the kites scattered like pigeons. When kites land in trees, the branch barely seems to register their weight. When the great eagle landed the tree practically bent double.

Kites, reintroduced by humans, don't have the osprey's fairy tale story. They don't have the size or the power of the eagle. But kites are special. People ought to know that.

The kite has a split personality. It is two birds rather than one. At first it is the lazy glider, sprawling on the breeze, and then, with the merest twist of the tail, it becomes a bomb, dropping silently towards earth, all that gangly frame drawn in, contracted, purposeful. How does it do that? How is it possible to change so dramatically?

They are fun. That's something else they have over their rivals. They like nothing more than to wait until another member of the pack has picked up some food, then chase and harry until they drop it. Then, such terrible pirates, they plunder the treasure. People often wonder why the kites would go to such lengths when there is other meat on the ground. I know the answer. They like to play.

We have no bird of prey as social as them, none as brilliant in colour, none as tolerant of us as they are. There are many ways to fall in love with raptors, but kites are the simplest, the gateway through which anyone can walk. They will dance in the sky, whether they have an audience or not.

But perhaps that is part of the problem. Nature is supposed to be about those moments when something appears unexpectedly and takes one's breath away. Kites don't adhere to this. Anyone can see them on any day of the year in the places they frequent. Do kites make it too easy?

These questions have long bothered me. Are the things that I love about the kites the same things that make other people indifferent? Often a visitor will tell me how, in their childhood, their parents would take them on holiday deep into the valleys of Wales in search of the red kite, that near-mythical bird living in exile there. Would people make such a pilgrimage to a feeding station now? Perhaps the thrill was all in the chase.

Could it even be that we at Argaty have failed in our duty of promoting the birds, of sharing the love? People sometimes refer to us as the hidden gem of Scottish birdwatching venues. To me the word 'hidden' takes the shine off the gem. For years we'd had this wonderful project, but we struggled to publicise it. In an average year we took around £15 a day. Subtract wages and we barely broke even.

A memory sticks in my mind. Some years ago we held an open day with activities in the visitor centre and free entry to the afternoon feeding session. The only people who came were a handful of volunteers and friends. We stood around sipping coffee and staring at our feet like mourners at a funeral, too embarrassed to talk or make eye contact. We had to do better than that. I felt as though we'd let Mike down. He had done so much for us, and for the estate. We should have been shouting his praises from the rooftops, but the farm took up so much of our time that we'd overlooked him and now he'd gone. And the kites? Perhaps we'd let them down too. In this bird we had something wonderful. It was time to let the world know.

In 2018, I began a campaign of almost shameless self-promotion. Every time we did anything remotely interesting I went online and told people about it. And I was amazed to find that the approach bore fruit.

In late January, journalist and presenter Euan McIlwraith visited the kites and recorded a lovely feature for Radio Scotland's *Out of Doors* programme. When a red squirrel was tragically run over on our road, I contacted a professional tree-climbing friend, Keith Burgoyne. He strung rope bridges high up in the canopy to allow the squirrels safe passage from one side to the other. This caught the attention of Andrew O'Donnell of Beluga Lagoon films. We began hanging around together, filming wildlife on the estate. In February, I ran a birdbox-making class, which quickly sold out. Another friend, Laurent, had to bail me out, spending hours helping to cut and drill wood in preparation. Buoyed by the success of the event, I ran further classes at local nurseries. More than fifty new nest boxes went up around Central Scotland that spring, thanks to our efforts.

I scheduled more events for the upcoming months, contacting charities and conservationists. Some knocked me back, some never replied, but many agreed to work with us. Our fortunes, it seemed, were improving. By making the runs, we were scoring a few goals.

I couldn't know it at that point, but life was about to get even better.

*

Below the now-ruined Argaty House are fields that I love more than any other. The long, wide plains are bookended on either side by beautiful mixed woodland. At the southern end lies a large pond. Swans nest there each spring. As kids, my sister and I used to fish there for perch, howling in confused anger as our hands were jagged when we returned our catch. Our parents neglected to tell us that the fish had spikes. More recently, Rowan and I found an old ruined rowboat submerged mere feet from the bank. It must have been there for decades, and yet I couldn't recall ever having seen it before.

What I love most about this part of Argaty is that the fields have trees in them. You don't see that often now. Take a walk in the countryside and you'll see what I mean. Trees have been marginalised, left to look on from the sidelines like the unpopular kids at the school dance. Rewind to the late 1950s. The war is over, rationing has at last come to an end and the government swears its people will never starve again. Trees are difficult to plough around and, in the quest to produce more food, agriculture must become more efficient. Many trees were removed from our nation's fields for this reason. A lot of our current wildlife problems can be traced back to that era.

Thankfully, these fields are different. Here oak, ash and lime have stood tall for centuries. The reason? For people's enjoyment. From the upper floors of Argaty House, residents could look down upon this large rolling expanse, watch the trees grow, see the ducks and swans come in to land on the pond. It must have been some view.

Two lime trees in particular are alive with activity. Cradled between trunk and branches are great balls of twigs around five metres wide. Jackdaw nests. Once, while gathering sheep, I saw a buzzard lift a chick from the largest nest. It took it on the volley, without breaking stride, and soared off into the distance, leaving the parents screaming hysterically at its shrinking shadow.

The jackdaws are always there, circling the territory in swirling, chattering packs. They began alarming the moment I set foot in the field one dank February day. The noise grew louder and more agitated the closer I got. Eventually they rose smouldering from the trees and drifted away like bonfire ash on the breeze. In the distance I heard a buzzard's plaintive mew. People sometimes confuse kite and buzzard calls, though to my ear they are very different sounds. Kites are joyful and shrill. They can't sing and they don't care; they screech away regardless. By comparison the buzzard's voice is mournful. Everything of their solitary nature is expressed in those notes.

Little mounds of earth were heaped across the field. Molehills. I found myself zigzagging to avoid them. The presence of moles is a clear indicator of the quality of the ground beneath you. If there are moles there must be worms for them to feed on; if there are worms the soil must be healthy. Worms feed on organic matter, digest it and break it down into nutritional forms that plants take up through their roots. Farmers often dig chunks of soil simply to count the wrigglers. The more they find, the healthier the ground, the more will grow from it.

On this ashen day, the ruin of Argaty House looked more ghostly than ever. I visited the house just once, when I was very young. The family who bought it from my grandparents invited us for tea. I was staggered by the size of everything, by the grandeur, by the thought that my dad had grown up there, surrounded by housekeepers, maids and nannies. I couldn't get my head around it, couldn't picture him living in such splendour. I still can't. Often I wonder what it was like for him, spending his adult life living next to, rather than in, his former home. Did he miss it terribly? When the fire tore through it, destroying everything inside, it must have broken his heart. It would mine. I've never broached the topic with him. Some sores are best left untouched.

Now the house resembles something from a horror film, a charred face with burnt-out eyes. In its dilapidated state, it is hard to recall how the mansion once looked. A picture of it does remain in my mind, but it grows fainter each year. One day I'll forget it entirely and see only the ruin.

On the far side of the fields is a long column of woodland known as the Westerton Strip. It marks the border of Argaty, where our estate ends. In previous years Mike spent a lot of time down here, setting traps for grey squirrels. These woods were a hot spot for them. A great number crossed onto our land from neighbouring estates.

'If you want to keep the reds, make sure you keep up the trapping,' he'd warned before leaving. 'The greys always come back. Especially in the Westerton Strip.'

Catching grey squirrels. That's what took me there that day. Leaping the fence, I walked to the middle of the woods, set three traps and quickly left.

Returning that afternoon, I repeated the same mantra over and over.

'Let them be empty. Let them be empty.'

I knew that I would shoot a grey squirrel if I caught it, but the thought of doing so gave me no pleasure. 'This,' I told myself, 'this right here is the reality of conservation.' People think you lead a charmed life. You look after animals, do the right thing, be the good guy. Sometimes that really is what the job is like, but not always. This can be a life of difficult decisions, a battle between what you want to do and what you have to. To live in this world, to really do your best by it, you need a little ice in your heart.

There was nothing in the first two traps. For a moment I thought I might get away with it, but that feeling quickly died. Further down the strip I could hear frantic metallic rattling. This sound did not belong in nature.

A dreadful feeling washed over me. It wasn't physical. I didn't feel sick, or faint, or anything like that. I just felt desperately flat. Owning this estate is such a privilege, but on very rare occasions I hate the burden it places upon me. Sometimes I wish there was someone else to deal with these things.

I could see the trap now. A small dark creature was darting from one end to the other. It never stopped.

I drew closer, and then my heart leapt. It wasn't a grey squirrel. It was a red.

A very angry red. Every time I tried to get my fingers through the bars and open the door my captive rushed at me. Using sticks, I finally pried it open. The squirrel sped up a tree, stopped on a branch twenty feet up and glared down at me, swishing its tail furiously. Heart still racing I sat back, looked up to the skies and laughed.

The following morning that sense of levity was gone. Back I came, pistol at the ready, once more convinced that there would be greys in the traps. Again I was wrong. Again, I'd caught a red. I couldn't be sure, but it looked like the same one.

A realisation hit me. This was what I'd been searching for. I broke into a sprint. Ten minutes later (wheezing, feeling sick, wishing I exercised more), I was back. With me I'd brought that trusted old ally: the ruined bird feeder.

10

Squirrels

Even in remote parts of the Highlands the kite . . . is
rare. An Argyllshire gentleman informs me that after a
residence of over ten years in that country he shot only
one Glead, and never saw its nest or heard of one being
taken in the country. Mr Thomas Edward, the Banffshire
naturalist, says, in Banffshire, where the Glead was
formerly plentiful, 'it is now rarely to be seen,' and Mr
McGilivray, a most accurate observer, tells us that so rare
is the Glead in the north of Scotland that in eight years
only one example has found its way into the hands of the
Edinburgh bird stuffers.

'The Birds of Glendale from the Sea to the Cheviots',
Alnwick Mercury, 14 July 1883

'That was a long walk,' Sarah said.

It was late. Several hours earlier we'd put Rowan to bed and I'd
left the house, intending merely to give the dog a quick walk then
come home. Instead I'd ended up in the woods again, checking
the stealth camera, gazing with delight at picture after picture of
red squirrels. The feeder was now so broken that the nuts could
be lifted whole from any of twenty gaps in the wires. Each time I
returned (and I returned a lot, during the day and at night) it was
empty.

'I really think I've found the spot,' I said.

And what a spot it was. Quiet and open, with a mix of conifers and broadleaves, it would certainly work for photographers. Often, approaching through the field, I'd see a squirrel hanging upside down on the feeder, gorging on nuts. Sometimes one would be chasing another. The camera was taking hundreds of photos per hour. In the end I turned it off. I had all the proof I needed. The squirrels had found the food and they were returning regularly.

Have you ever had that feeling when you know something really positive is about to happen? The sensation doesn't come often. In my working life I'd only had it a couple of times at the most. I had it now.

Within days, Marion and Fiona, two photographer friends, had captured pictures of the squirrels. To my knowledge they were the first ever taken of red squirrels on Argaty.

We set up a working group. Marion, Fiona and her husband, Alan, met Mum down in the woods. Perches were strategically arranged, a variety of food was placed upon them and a pop-up hide was erected in the feeding area. The squirrels couldn't have cared less about the tent, or our presence within it. There's something very cat-like about their attitude towards people. I wouldn't ever say they grew to like us, even though we were their source of food. At best we were tolerated; at worst, scorned. I remember turning up late with their nuts one afternoon. I'd been distracted by some other task and arrived to find the feeder empty. There came a wittering sound overhead, a gobbling, like that of a turkey or a hen. I realise now that I'd heard the sound so many times before and had always written it off as some strange birdsong. That day I looked up. On a branch overhead was a squirrel. I'd never seen anything so comically angry. Had it spoken English, I'm sure that every second word would have been an expletive.

And all of this was directed at me! His guardian, his benefactor. I couldn't believe it.

Each week we saw more squirrels. Some were dark, some bright. Some were tolerant, some territorial tyrants who tore after the others in a high-octane game of Follow the Leader, trying to defend their spot.

Once I entered the woods and saw that the tent was vibrating. Teenagers, I thought, with alarm. But all the way out here? I coughed to announce my presence and a squirrel burst out, swearing at me as usual. A valuable lesson was learned that day. Where food is involved, squirrels can be very determined. A box of nuts may appear tightly sealed, but a squirrel can disabuse you of that notion. Sights like this made my day. Every time I saw these creatures I learned a little bit more and my appreciation of them grew.

It dawned on me that, for the first time since Mike's departure, work was truly going well. That felt a little strange. Part of me knew I'd cheated my way to victory. When I first began looking for squirrels, I thought I'd develop some form of fieldcraft which would lead me to them. Instead, I'd stumbled across them by accident. It felt a bit too easy. A silly thought, perhaps. More serious was the realisation that if this was to work out, and it felt very much as if it would, I would take the credit for someone else's work. Mike had culled the greys, opening the door to the reds. He'd played the defence-splitting pass, leaving me with an open goal. I was set to tap it in and celebrate without him. That still troubles me.

Nonetheless, I was happy. I couldn't claim otherwise. Having stated my intention to do this, I felt great satisfaction to see it coming to fruition.

A variety of birds soon found the food. It was fascinating to watch them. The greater spotted woodpecker flew along the edge of the treeline, his bobbing flight so distinctive that now when I see a bird flying as if riding a wave I know it instantly. I watched him drum, always on the same tree, the one with the best acoustics, marking his territory and advertising himself to potential mates. It wasn't long before he found one and they began patrolling the site together. They were the lords of this jungle. Everything else bowed down before them. You always knew when a woodpecker was around because the smaller birds hung back, waiting respectfully for their turn to feed.

Second in line to the throne were the nuthatches. These little brawlers, with their black mascara stripe, grey cloak and orange bib, are still quite the novelty here. I'd spotted my first one on Argaty a few years earlier and had had to reach for the bird book in order to identify it. It transpired that nuthatches are relatively new visitors to Scotland (though their numbers have long been plentiful south of the border). The first nesting pair up here was recorded in 1989 and since then they've expanded their range northwards. This, experts believe, is partly due to climate change, partly due to people feeding birds in their gardens.

Nuthatches are interesting birds, able to walk both up and down a tree trunk with ease. Their movements are not unlike a squirrel's. When either descends headfirst towards the feeder, they arch their necks, gazing all around like acrobats dangling from a trapeze.

I used to think that the name nuthatch derived from their habit of stashing nuts for the winter. Did they store them away and wait for them to ripen? In hindsight I wonder why I would ever have presumed this. After all, why would anyone use a term like hatching to describe the ripening of a nut? Nuts, quite clearly, do not hatch. In fact, the name has evolved from the Middle English word *nuthak*, meaning hacker of nuts. When topping up the feeders one morning, I heard this bizarre tapping, cracking noise.

I knew it wasn't the woodpeckers; it didn't have their regularity or rhythm. Then I saw it. The resourceful nuthatch had taken a hazelnut I'd laid out for the squirrels, jammed it into the crevice of a tree and begun hacking at it with its beak, chiselling through the shell.

We saw so much more in the months that followed. A flock of long-tailed tits appeared for a spell. They arrived, fed and moved on. Why did they leave when there was so much food here? One of nature's mysteries, I suppose, but I was sorry to see them go. They huddle together at night to keep warm and, in the breeding season, help other parents in the flock to raise their young, should their own nest fail. What lovely traits. I'd have enjoyed seeing that.

Chaffinches and tits became regulars. Blackbirds sang their throaty song. Robins loitered below the feeders, hoovering up the scraps, challenging everyone to a square go. Jays were seldom seen, but their shearing metal voices were always heard. The treecreeper pottered up the conifers, white tucker on, little hooked snout digging for dinner between the plates. Deer wandered past, unaware they were being watched. Voles and wood mice ping-ponged at incredible speed between all the perches, swiping food and bolting. There were many characters to watch. The wood holds a lot of secrets. If you sit and look for long enough, you might be lucky enough to learn some of them.

The biggest surprise was always the sparrowhawk. He flipped through the trees from time to time, chainmail-armoured chest glinting in the sun. Sensing the danger before he'd even arrived, the birds alarmed and scattered. The squirrels ran for cover, too. I was always the last to react, always the one blown away by the raptor's arrival. Animals make us look very slow sometimes.

The photographers and I debated feeding the hawk. It would, no doubt, have been a big attraction, but it seemed to go against our aims. We hoped to feed the squirrels, and by doing so help re-establish their population here. I know that everything has a right to eat, and if the sparrowhawk ever caught a squirrel I'd be fine with that. I wouldn't want to be the one responsible for this, however, and deliberately feeding predator and prey in the same spot was risky. The idea was jettisoned.

After much experimentation we decided upon a position for a hide and I placed an order with a shed manufacturer. In April, they would come out and build it.

As that day drew nearer, I grew increasingly anxious. This close to the goal I feared that things would fall apart, that the squirrels would be predated or else emulate the pine martens and suddenly, inexplicably, vanish. My photographer friends kept me sane, spending hours in the woods, reporting back each time they saw a squirrel. How grateful I was for that. Embarking upon this new way of life, I had hoped to develop a wide circle of enthusiastic and proactive friends who believed in the same things we did about how the countryside should be managed. In these people, I had found that.

11

Spring

We have . . . testimony of their presence in our area, viz.,
July 4th 1884, but for obvious reasons we prefer not to
state our authority, nor indicate a closer locality . . . At
Inveraray the Kite was abundant about 1841–45, His
Grace the Duke of Argyll having himself seen nine at
one time sailing round the castle. They have disappeared
absolutely, all having been killed off by keepers.

J. A. Harvie-Brown & T. E. Buckley, *A Vertebrate
Fauna of Argyll and the Inner Hebrides*, 1892

The year does not begin when January deposes December. No,
after months of steady negotiation the new year walks in gradually, when it is ready, one small step at a time.

The snowdrops are the first indication of change. An army of
milky white heads pierces the cold ground in February. Daffodil
shoots appear next. Then the flowers reveal their faces. At night
the tawnies renew their vows, the female's 'kerwick' answered by
the male's 'whooo'. The cawing ravens follow suit. From the wild
fuchsia at my back door, a proud little robin sings its sweet mating
song. Soon all of the small birds are darting into hedges and gaps in
dykes, carrying moss, grass and nesting material in their beaks. As
always, the biggest explosion of the floral firework display comes
last. Bluebells shoot up, past leaves and fallen branches. When

their flowers are in bloom it is as though a sea of blue has washed over the once-muddy forest floor. I see them and know that spring is here, the old year has gone, the new year begun.

Spring took a long time to come in 2018. Winter kept walking out, then, as if unable to step away from an argument, coming back in again. The weather flummoxed the kites. In the mild temperatures they began to flirt, flying in tandem through the sky, racing in each other's slipstream. Then came another chill, and survival seemed more important than sex. At last the thaw really did come and they resumed their wooing.

One of the most dramatic sights you can see in nature is two rival kites talon-locking. From up in the clouds the birds come together, take one another by the hand, then plummet, spinning and tumbling, gaining speed as they hurtle towards the ground. In the final second, when impact seems inescapable, they release their grip, separate and fly off. How this doesn't lead to an accident is beyond me. It reminds me, strangely, of a scene from my childhood. Like many unfortunate kids across Scotland, every year my teachers inflicted a term of country dancing on us. Two memories linger from those awful days. The first is that I always ended up dancing with a boy. It was either that or dance with a girl, and as a young lad the very idea of doing *that* was disgusting. My second memory is that at some stage I'd be required to spin my partner. That's when the dance became a blood sport. The spinning grew faster and faster, deteriorating (inevitably) into a contest to see who would be first to fling their other half into oblivion. It's that moment – the point just *before* the carnage – that I think of when I see two kites locking talons.

We'd seen less of the squirrels during the colder periods. Minus temperatures rapidly deplete their energy, so they retreat to their dreys, use their tails as blankets and simply try to stay warm. Sometimes these little squabblers, at war for so much of the year, will even deign to share a nest. Two tails, it seems, are better than one. So rarely do they venture out on such days that it's small wonder so many people believe that they hibernate in winter. With the thaw they reappeared and, like boxers training for a fight, began a rigorous routine of bulking up, building reserves for the mating season. They sat inside the feeder boxes for half an hour, sometimes longer, devouring more nuts and berries than seemed possible in creatures so small. Whenever an unsuspecting rival opened the lid an almighty chase began. The dominant squirrel was remarkably dogged in these pursuits. The intruder would try all sorts of evasive manoeuvres, circling trees, leaping from branch to branch, hiding in the canopy, but to no avail. A squirrel chase only ends when the interloper flees.

Weeks passed, winter faded out of memory and, as spring found its feet, tourists returned to Scotland. I was relieved to see them. Only when they leave the country over autumn and winter do I realise how much I thrive on their company, their enjoyment of the kites and their enthusiasm for the project. Their return is one of the things that makes spring special – but it is only one. Over the many seasons of rebirth I've enjoyed at Argaty, we have seen such great things.

There was the original reintroduced kite who lived such a long, ferocious life that by the age of fifteen she'd been through four husbands. Usually kites mate for life. Not her. We named her Elizabeth Taylor.

Then there were the CCTV kites. In 2011 Mike had two cameras installed in the woods. One was focused on a kite's nest, the other a buzzard's. Setting this up had been tricky. These were the days before wireless cameras were readily available. He'd had to string cables from the nest tree, through the woods and down to our visitor centre several hundred metres away. We enjoyed great views of two chicks hatching and subsequently growing.

Sometimes, when Mike was feeding the kites I'd stand waiting in the yard until they dived, then run to the visitor centre and watch as the parents brought the food we'd provided back to the nest. Those were magical moments. A few short weeks after these particular chicks had hatched, a terrible storm hit. Gale force winds ripped through the region. Mum, Mike and I stood watching the monitor in horror. The tree was swaying violently. We caught a glimpse of the chicks clinging on for dear life as the trunk swung from left to right, then the screen went blank. We all agreed there was no way the birds would survive. To have watched them hatch and grow, seen their fluffy white heads craning skywards as their parents presented them with food, made their certain deaths very hard to bear.

The storm abated and Mike ventured back into the woods, expecting the worst. Would he find the chicks dead at the foot of the nest? Perhaps the foxes would have them already. The view on his approach did little to inspire confidence. Giant root plates towered up into the sky. All around trees had dropped like dominoes.

Miraculously, however, both the nest tree and the camera tree remained upright, though the camera cable had been torn loose.

The chicks were okay.

In mid-June we ringed and tagged them and that should have been the end of the story. The birds should have waited a few more weeks, then fledged the nest. One, however, had other ideas.

Turning the CCTV on a week later, we were met with the most alarming sight. One of the chicks, fitted with wing-tags reading S4, was not there! Mike hurried back out. There was no sign of her below the tree, no sign on the nest either. Then a scrabbling sound met his ears. Clinging to a tiny branch just beneath the nest was S4. An SOS was sent to Keith Burgoyne, our resident treeclimber, who sped over, shot up the tree and returned her to safety.

Both chicks fledged successfully, although S4 later found herself in the wars again.

Marion, our photographer friend, had recently joined us as a volunteer ranger. On one of her first shifts in the hide S4 flew lopsidedly out towards the feeding area, one wing held aloft, the other drooping. Marion called Mike, who caught S4 and delivered her to the SSPCA. Fortunately the wing was only sprained and after two weeks' rehab she returned to Argaty for good.

Her parents changed nests each of the two following springs, forcing further camera relocations. Once, when spring had rolled into summer and the chicks had fledged, Mike and I went to the woods to wind up the cables for another year. I was looking up, trying to spot the nest, when I almost trod on a chick lying motionless but utterly unharmed on the ground.

What it was doing there was unclear. Perhaps it had fledged before it was really ready to fly. While Mike went to fetch Lawrence, a climber who lived on the farm, I sat and watched the bird. It remained still, eyes wide and beak agape, ignoring the midges crawling over its mottled feathers, the flies swarming around its head. This bird really was an insect magnet. (In the few minutes we spent together it transferred some of its friends to me. Sarah was less than impressed when fleas subsequently broke out in our home and the place had to be fumigated.) The chick was taken back to the nest and we were hugely relieved when it fledged successfully soon afterwards. Sadly the storms hit every year of the CCTV cameras' short lives and an absolute howler of a gale eventually trashed the cables, bringing an end to our nest-watching.

By then, however, we had other excitements to show people. Mike had sent Keith and Lawrence up trees near the visitor centre with chainsaws, wooden pallets, bags of sticks and turf to build giant artificial nests, hoping to attract ospreys.

'I'll be dead before an osprey comes here,' said my dad, rolling his eyes.

'Erm . . . what I meant was I'll be dead before ospreys actually *nest* here,' he said, backtracking swiftly, when just a few weeks later a male osprey landed on one nest and began to build it up with sod and twigs.

Fortunately for Dad, predictions of his own premature demise proved to be inaccurate, for the osprey did not nest there. When he was away fetching more sticks, a pair of kites stole in and claimed the nest. They raised two chicks that year, the first kites ever to have fledged from an artificial osprey nest in the UK.

Stories like these make spring special. So many unexpected things happen in this season. Each year brings something different. Truly, there can be no finer time of year.

12

Nesting Season

In 1892 Harvie-Brown visited the ancient site of a veritable Goshawk's nest in Rothiemurchas and Glenmore ... What now remains is a mere stump, three or four feet high, similar to that in which on the same day we took a Crested Tit's egg. Since its early occupation by the Goshawks it was tenanted by Kites. The then keeper (some thirty years ago) having set a trap and caught the Kite, forgot to look at his trap, and an old Spey fisherman found it hanging there, and, as he could not climb, he got an axe and cut it down, in order to secure the Kite's feathers for salmon flies – so runs the story.

J.A. Harvie-Brown & T.E. Buckley, *A Vertebrate Fauna of The Moray Basin*, Vol. II 1895

With the new season came a new challenge. Spring is the busiest period in the raptor worker's year. A race begins in March, gathers pace in April and ends in a madcap sprint to the finish in June and July. The aim is to find nests, monitor them to see if they're in use and ultimately remove the chicks, fit them with a leg ring and return them minutes later.

It's a crazy period, with start and end points that nature alone dictates. From the moment the birds start their mating displays, raptor workers must be on the ground, searching.

Were it just kites that we monitored on Argaty, things would be rather less manic, but instead we try to ring every young raptor, raven or owl that we can find. To make things even harder, they all nest at different times. Typically ravens and tawny owls start early, meaning that their chicks will be the first to be ringed. Kestrels are usually next, though this is variable. Barn owls must be checked constantly, as they can nest at any point. Kites and buzzards come later and we have them in abundance. When their chicks are ready to ring we run from one nest to the next, cramming as much as we can into the time available; there are never enough hours in the day.

In addition to fitting the chicks with rings, we also weigh and measure them. For kites in particular, this is important. Males and females have the same markings and colourings. The best means of ascertaining their gender is by weight. While there can be some overlap at ringing time between a heavy male and a light female, by the time they leave the nest the girls should be marginally heavier than the boys. Of the various theories advanced on this a likely explanation is that they have evolved this way due to the demands that life will place upon them. Females will incubate eggs, males will hunt. She must be bigger, with greater reserves, he quicker and more agile. In some species, the size difference is great, but in kites it is scarcely noticeable. (Females tend to weigh in at 1–1.2kg, males 850g–1kg.) The probable reason for the similarity of size is that the kites' main food source is carrion. If they hunt at all it tends to be for small, slow-moving prey. Smaller raptors – male sparrowhawks, for example – need to be lightning quick to catch their food; kites do not.

The purpose of nest-monitoring and leg-ringing is to keep records, and in this job records are crucial. The information we collect is published each year as part of the Scottish Raptor Monitoring Scheme report. Our sample data, collected from thousands of raptor territories, tells us how well, or otherwise, each species of bird has fared in the breeding season. It is used to inform policies from local to national level and has been cited in hundreds of scientific papers. Raptor monitoring is vital work, by parts enthralling, upsetting and exhausting, but at all times essential if we wish to know anything about our birdlife.

For years Mike monitored the nests on Argaty. When he moved south, I applied successfully to join the Central Scotland Raptor Study Group and cover the patch. My first group meeting, in the autumn of 2017, had proved quite a surreal experience. There I sat, the inexperienced new boy, surrounded by many of Scotland's most knowledgeable raptor workers. I felt like a child allowed to sit at the grown-ups' table. This sensation was enhanced by the fact that several of those gathered there that night had been members of the RSPB's red kite reintroduction team. Duncan Orr-Ewing, Kevin Duffy, Simon Zisman … When I was young, these men used to sit around my parents' table talking about kites. I'd grunt an awkward teenage greeting at them as I ran out of the house after school to kick a football in the yard.

During the meeting, the members spoke with incredible knowledge about the birds on their patch. They seemed to spend most of their lives in the field. This task would be bigger than I'd anticipated.

Now spring was upon me and I was gripped by an inescapable fear. This was not just a big task, it was huge. Nearly a quarter of Argaty's 1,400 acres is woodland. That's one hell of a patch for a novice to cover. How would I find all the nests? Would I find any? Visions of nil returns haunted me. Every time I thought of it I felt a little bit sick.

Just as my anxiety was reaching boiling point, help arrived. Keith Burgoyne offered to mentor me. A lean, lively character, all energy and wiry strength, Keith is a seasoned raptor worker

A red kite swoops for food.

Sorting the meat from the grass. Sometimes kites grab more than they bargain for!

Mum and Dad watch the kites at feeding time.
(Andrew O'Donnell)

Mike McDonnell and visitors watch the kites from the hide at Argaty.

Argaty House in its former glory, circa 1920s.

Mike with a barn owl chick. Mike did so much to bring wildlife back to Argaty.

Mike's bathroom cabinet nest box: An unorthodox design, but the starlings love it.

Me with tawny owl chicks, circa 2010.

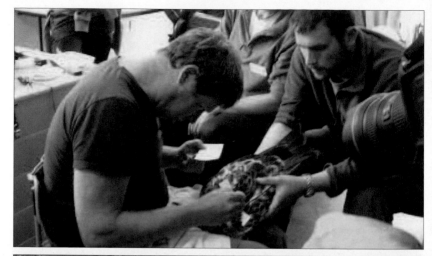

Top. Former red kite officer Kevin Duffy fits wing tags to the young kite that was shot in 2009. Moments later the falconer Stewart Miller released it on Argaty. (Marian White)

Above. Red kite chicks in the osprey platform. These were the first ever kite chicks to fledge an artificial osprey nest in the UK.
(Lawrence Monckton)

Right. Battling against the elements: A kite braves the weather in search of food. (Clare Harte)

Red kites and buzzards feeding in the winter. Cold spells are hard for these birds. They are always glad for our food.

Red kites gather for the winter roost. (Philip Price)

Sarah and Rowan en route to the squirrel hide. (Tom Bowser)

In winter, red squirrels grow delightful ear tufts. Those on this particular squirrel were quite amazing! (Robb Allen)

A red squirrel on the prowl for food.

In the woods with Keith Burgoyne. (Ràna Webster)

Left. Gill the otter as a cub. Colin Seddon and his team did amazing work to get her back into the wild. (Colin Seddon)

Below. Gill enjoys a bath at the SSPCA's rescue centre. (Colin Seddon)

Bottom. Releasing Gill at Argaty. (Colin Seddon)

Our volunteer Sandra McDerment goes to war on the overgrown bumblebee garden. (Ràna Webster)

Dave Anderson fits leg rings to one of our osprey chicks.

Our female osprey and her young wait for the male to return with food. (Nest camera footage)

Beavers are one of Scotland's great rewilding successes. If not for the avian reintroductions, would we have beavers back? (Philip Price)

Beavers are now in many of Scotland's rivers. One day they'll reach Argaty. We can't wait. (Philip Price)

'Strike a pose': A kite takes to the sky with a fistful of food.
(Philip Price)

Duncan Orr-Ewing satellite-tagging red kites in Extremadura, Spain, in 2019.

Winning the 2018 Nature Tourism Award at the RSPB's Nature of Scotland Awards, with Euan McIlwraith and Kate Humble. We are still stunned to have won. (Simon Williams)

Rowan spreading poppy seeds.

Mike's wildflower garden inspired me to sow other wildflower plots on the farm. The areas that flourished have provided essential habitat for wildlife. (Marian White)

Heavy-lifting time: Our wonderful volunteers Talia and Richard go to work on the raised-bed pond. (Ràna Webster)

Note to self: Never start a construction job in winter. By the time we'd finished work, this ground had become a mud bath! (Ràna Webster)

Getting up close to a common blue damselfly. (Philip Price)

An emerald damselfly perches pond side. (Philip Price)

Above. The wonders of the underwater world: The raised-bed pond became an amazing home to wildlife. (Ràna Webster)

Right. Heading off to feed the kites with Rowan.

Me feeding the kites.

A wing-tagged kite dives for lunch. Wing-tags taught us so much about these birds' lives.

With their flaming feathers, snow white heads and piercing pale eyes, kites may be Britain's most beautiful raptor.

Food fight: In recent times the kites have become increasingly territorial, particularly where food is concerned. (Alan Jones)

and an expert tree climber. It was he who rescued our nest camera chick, who installed our osprey platforms and our red squirrel rope bridges. In spring and summer he is out each day studying every type of Scottish raptor, both in Central Scotland and in the Pentland Hills near Edinburgh. The self-taught naturalist is a rare breed these days, but Keith is one of them.

In his youth he would skip school, spending days in the country-side, discovering its ways by simply sitting and watching it unfold. A natural showman who loves a good story, he makes for great company. Every day with him is an education. You don't indulge in small talk; you just ask questions, listen and learn.

In March, while the kites played chase overhead, he and I set to work. How we walked in the weeks that followed! Through field and woodland we went, covering miles. We were soaked by rain, soaked in sweat, soaked by rain again. Occasionally the terrain was easy, but mostly it was the opposite. Your feet disappeared into hidden bogs, marshy water filled your boots, fallen trees blocked your path, branches scraped your skin. The days drained you and by night you were fit for nothing but sleep.

Keith has a distinctive way of walking. Hands knitted behind his back he'll stop suddenly, inspecting something at ground level, a drill sergeant noticing an unpolished boot in the platoon. On those walks he saw everything and, once he'd shown me, I began to see things for myself.

The cluster of grey feathers on the ground belonged to a pigeon. A raptor killed it. How did we know? Because the quills were intact. Raptors pluck the feathers from the body before eating. The unfussy fox bites feather and meat, leaving a pile of snapped quills in his wake.

The claw marks on that tree were made by a pine marten. Search nearby and you'll find twisted brown scat (faeces) on a dry-stone wall. The marten was marking its territory.

And the clear, gelatinous goop on the ground? A pregnant frog had its stomach slit. Probably the work of a buzzard or sparrow-hawk. Maybe a goshawk. Those tiny black dots in the slime were the frog's eggs.

There was so much to see when someone showed you how to look.

The aim in the early part of the raptor season is to try and locate last year's nests. This is not an easy task, particularly in the conif-erous woods where trees keep their needles year round. Things become even harder when the deciduous trees come into leaf again, so the key is to find nests early, while they remain visible. Some birds – kites and buzzards, for example – help you. In early spring, pairs circle over their nesting site, calling to each other. This act is known as displaying. Should you happen to see this, or later chance to see a bird flying into the woods carrying twigs or other nest-building materials, you have a head start in your search, a rough idea of where you should be looking. By the second week of April the kites will be on eggs. Miss your chance before that and your only option is to do it the hard way: walk through every wood, looking up, hoping to spot them.

That spring, Keith and I did it the hard way. Finding a nest in this manner was essentially a problem-solving mission. We searched for flight paths (clearings wide enough for a raptor to fly through). We turned constantly, double-checking everything because a nest might be visible from one angle but utterly hidden from another. We listened for the birds' calls, and scanned the ground for feathers, droppings or anything that might indicate that a nest was overhead. Sometimes there was nothing up there. The feather on the ground merely indicated that a bird liked to perch above. Sometimes there were too many droppings, sug-gesting that a communal group of birds passed through, starlings perhaps. Sometimes, however, that clue was our golden ticket.

Gazing up, we'd see our goal: a basin of twigs up in the heavens, perched somewhere between the trunk and branch.

Once we'd found the nest, the next step was to monitor it. Over the following weeks we would return again and again, checking to ensure that it remained in use. The nest can fail due to bad weather, disturbance or predation; the pair can begin building then decide against using the site and move somewhere else, as had happened each year with those CCTV kites. The picture is constantly changing. There are two ways to tell whether or not a nest is still active. Sometimes it's possible to find a vantage point where the chicks are visible, but, if not, streaks of white excrement at the foot of the tree are your best guide. Although the nest may be filled with all sorts of rotting meat, young raptors have strange double standards when it comes to hygiene. They never shit on their own doorsteps, opting instead to reverse their fluffy bottoms to the edge of the nest and defecate onto the ground. If there is no tell-tale white below the tree, the nest has most likely failed or been abandoned. On each visit I kept my fingers crossed, praying we'd find fresh splatter.

We ringed the raven and tawny chicks in May. The ravens were docile, lying there with massive beaks gaping. Though little more than wee white pom-poms, the tawnies clicked their beaks angrily, letting us know they would take us in a fight.

Something was changing in me. In years gone past I would often join Mike and Keith when it came time to ring the chicks. While I enjoyed the experience, I couldn't quite share their levels of excitement. For them, ringing was the culmination of weeks of work. For me, it was something altogether more fleeting. I had come in at the closing act, the end of the race. This year was different. I was delighted by these little birds, captivated by their very otherness, thrilled to be ringing them after weeks of monitoring. I felt a connection to them that had been lacking in previous years.

Mike and Keith were addicted to this work, and now I too had caught the bug.

*

In mid-April the men had come to build our hide. A month later, satisfied that the squirrels and birds were used to the new structure, we opened to the public. To celebrate, our photographer friends Alan, Fiona and Marion came for dinner at my parents' house. Although it was a happy occasion, something was nagging me. I'd felt it more and more as the arrival of the hide had drawn closer. Now it was tugging on my sleeve again, refusing to be ignored. I looked around the table. What warmth I felt for these people. What gratitude. For months we'd been working together. We spent hours with one another and, even when apart, we texted or emailed. Many nights I'd lie awake until the small hours, writing messages to them while Sarah and Rowan slept.

I am not someone who deals well with change. I tend to wallow when I think of happy times now passed. That evening, sitting with my friends, I was very aware of this. We had opened this hide, handed it over to the world. Our first visitor was due in a matter of days. A period of great collaboration was coming to an end.

There was, as I could see, only one solution to the problem, and in a quiet moment that evening Fiona articulated what I had been pondering.

'So . . .' she said.

'So?' I replied.

'What next, then?'

What, indeed? My outlook and ambitions had changed since the previous summer. Back then, the squirrels were the sole plan. Nothing lay beyond. I didn't even know why I'd alighted on the idea of feeding them, except that I thought I might manage it, because I wanted Mike to stay and because, for still unarticulated reasons, I desperately wanted to stop farming. I had had many other ideas in that period, from renovating barns for never-named purposes to setting up soft-play centres or petting zoos. Of these, the squirrels were the only one that really appealed and therefore the only one that I had pursued with anything approaching conviction. But still, the plan had set limits. In my mind the farm remained something of a fixed entity. I could play around the edges of it, find unused spots and work on little schemes like the squirrel feeding, but the idea of doing something on a larger scale, making conservation the core part of our lives rather than an interesting side project, had not been on the agenda.

Through the pursuit of the squirrels, however, I had learned a lot about the place and about myself. I knew more about the wild creatures that were out there and, thanks to the nesting season, I now knew where some could be found. I knew the thrill of setting something up, of putting food and a camera out in a new spot and hoping an animal came to it. I knew that, for me, driving towards something is somehow more exciting than reaching the destination. And I knew that I wanted to work with my friends on other projects. This could not be the end of something, it had to be the start.

'You know,' Fiona said, 'you could have lots of hides here if you wanted them.'

I nodded. 'It could be a world-class wildlife destination.'

I had no idea how anyone would react to this statement. To my relief, nobody laughed, nobody looked doubtful.

Future possibilities dominated the rest of the conversation that evening.

13

The Greedy Gled Fallacy

Within the past thirty-five years the Kite has become extremely scarce within 'Dee' and it is now very doubtful if a single pair is to be found breeding within the district. In 1865, six were seen by Mr A. Gill nailed to a door at Monaltrie, Ballater; and in 1867, I saw five with Mr Brown, Micras, Near Aberdeen, all locally-killed specimens. In the *Aberdeen Free Press* of December 13th, 1872, it is reported that a Kite was killed at Balmoral in July 1871 ... The tail feathers of the Kite were much in repute for the dressing of artificial flies; indeed few flies were considered equal to that made from the tail of the 'Glen Tanner Glead' by Dee anglers, and to this fact, we presume, the extinction of the bird is in great measure due; for that it is on the verge of extinction, if not actually so, is beyond question.

G. Sim, *A Vertebrate Fauna of Dee*, 1903

In the early part of the breeding season male kites are the sole hunter-gatherers. While their wives sit warming the eggs, the menfolk forage. We see them only infrequently during that period. They appear at random, quickly take some meat, then vanish for long spells. To the untrained eye it might look as if they've lost their taste for our food.

In reality, they have other things on their minds. The male kite is an insecure, untrusting husband. His worst nightmare? That while he is away some rival steals in, takes his woman, fertilises her eggs. So each time he brings food home he copulates with her. Meanwhile, in the fields around Argaty, lambing is in full swing. Some think that lambs arrive at random at the start of spring, but it is actually a carefully planned process. The average gestation period for sheep is 147 days. We set the rams loose with the ewes for a short spell in the last days of November, planning to begin lambing at the end of April when the weather has improved and the grass has begun to grow. When the lambs are born, the kites are presented with plenty of free food. Though the lambs themselves are safe from harm, the ewes leave placenta in the fields which the birds gladly accept. With sex on the brain and free meals all around, it is no wonder that we see so little of them in the first days of May.

All of this changes with the arrival of the chicks. By this stage lambing is all but over. Those few ewes left to lamb are moved to the shed, where they can be easily watched. With young to provide for and no more placenta to find in the fields, the male kite returns to our food.

What a different bird he is. It is as if some radical transformation occurred in the weeks he was away. What little timidity he once had is gone. Heedless of any danger, he dives instantly the moment we put the food out.

The spring of 2018 slid so subtly into summer that nobody noticed the change. For weeks a vast area of high pressure had sat smothering Europe. Temperatures were rocketing. The heatwave was killing the land. The ground turned hard as rock, the farm tracks died of thirst, coughing up dust-bowl clouds in their final parched breaths. On a forest road frogs had laid their eggs in a puddle. Keith and I spotted the tadpoles swimming in an ever shrinking pool. We did all we could, topping the puddle up with drums of water, but it was already too late for many of them. Their ink stain corpses lay cooking on the ground in the places where water should have been. Climate change was biting with an intensity that was impossible to ignore.

By the early days of June, with the kite chicks feathered enough to stay warm without brooding, large enough to be safe from avian predators, their mothers joined in the search for food. Small prey such as worms, an appetising meal in earlier days, were no longer of any use to the nestlings. They were bigger now, and hungrier. The females looked ragged when they took to the skies again. Those early stages of incubation presented a perfect opportunity to moult old feathers, but replacements had yet to grow in. They flew regardless, driven to provide.

This determination was quite startling to witness. One day in mid-June I walked out to the field with the bucket of chopped meat. Tilting my head upwards, I sought out the kites above. They'd climbed high that morning and had sat coasting on the thermals ever since, watching everything. In that clear weather they could see for miles and must have picked me out the moment I left the hide. Recently they'd begun diving when I'd retreated little more than a metre, a clear sign that their chicks were hungry. I sped up, keen to get the food out and move away before the onslaught began.

At the feeding area I upended the bucket. One measly venison chunk flopped to the grass. The rest had dried out in this relentless heat and was now stuck to the plastic in some thick, rank glue. I was tipping the bucket, banging the bottom, watching the paltry picnic slowly slop out when a kite dived straight over my shoulder. It moved so quickly that I didn't even see it, but I heard it and felt it. For a mere moment, the air around me took on a new tone and tempo.

Returning to the hide, I was greeted by a chorus of laughter.

'That was like a scene from Hitchcock's *The Birds*,' one visitor said.

'It nearly took your head off,' said another.

I had to laugh, too. It had been an incredibly close encounter. But later that day I was struck by a worrying thought. What had the visitors taken from their visit? Did they imagine that kites were always like this?

How easy it would be to form that impression, to see a fearless, reckless bird and forget those hungry beaks in the nest. I thought of those people of past centuries, those who killed kites in their thousands, who wiped them out. And for the first time in my life I thought I understood why they believed in the legend of that terrible killer, the greedy gled.

*

In my early years working on the farm Dad drove a Ford pick-up truck. An image of the agricultural equipment he sold was branded on one side of the cab, a picture of a kite was on the other. Often I'd deliver the equipment to customers. As I drove into their yards, they'd eye the kite with suspicion, cast a worried glance towards the lambs in the field or the pheasants in the pen. Then came the standard comment: 'Hope those bloody things don't come here.'

The fear of the greedy gled lingers on in some quarters still. Of all the tall tales that our predecessors handed down, this one has proved particularly resistant to correction.

Why these sentiments should have such a lasting legacy is not immediately obvious. It is not as though we always hated these birds. The evidence suggests that they were both loved and loathed historically, that our feelings waxed and waned over time and that geography, specifically whether one lived in the town or the countryside, often proved a determining factor in one's disposition towards them. From the fifteenth to the sixteenth century, kites were protected by law in England and Wales in recognition of their work cleaning carrion from city streets. Although such protection never existed in Scotland, the fact that we named towns after them suggests some degree of affection. After all, nobody has ever named a settlement after an animal they considered a pest. While they must have had their admirers in towns and cities, kites were probably less popular in rural areas. Starvation was an ever-present danger for impoverished country dwellers. Survival depended, amongst other things, upon one's ability to hunt small prey and steward free-range poultry. Protecting these animals from predation by kites must also have been essential. As we began to clean up our filthy city streets, kites must have caused country folk increasing difficulties. With less carrion available, the birds began to move out of the cities and into the countryside, where they had never been wanted. Perhaps this is where our fondness for them truly began to wane. The Tudor vermin acts of 1532 and 1566 would follow. The mass killing of kites in England would begin.

By the mid-1800s much of rural Britain had changed. Agriculture had become more efficient and our farmers' output was now amongst the highest in the world. This did not lead to a lessening of tensions between people and predators, however. For landowners, sheep-farming had become big business. Threats to profit would not have been tolerated. Meanwhile, for those impoverished rural people, the threat of starvation remained very real; famines in Ireland and the Scottish Highlands in the 1840s were a powerful reminder of that. For those people the idea of surrendering any food supplies to a predator would still have been unthinkable.

The idea that gleds were voracious killers was now so prevalent that it had entered daily discourse. In the 1876 *Scottish Gallovidian*

Encyclopaedia, John McTaggart wrote 'We say of anything that has got into greedy keeping, that it has got into the *gled's claws*, where it will be kept until it be savagely devoured.'

This idea is reinforced in a 1890 edition of the *Newcastle Courant* in which the rules of the Scottish children's game Shue-gled-wylie (shoo wily gled) are explained:

> Shue-gled-wylie is a game in which the strongest acts as
> the gled or kite, and the next in strength as the mother of
> a brood of birds, those under her protection remaining
> behind her, one holding the back of the other. The gled
> tries to seize the last, while the mother cries shue, shue
> and endeavours, with extended arms, to ward him off.
> Should the gled catch all the birds he wins the game.

The kite, so these sources would have us believe, was a rapacious bird, a sharp-clawed predator, a bully that preyed on the weak. That the acts of plunder only occurred in spring, when an abundance of small prey was available and the kites had young to feed, was not considered.

As we know, by the mid-1800s the countryside had changed in another fundamental way. Game preservation had grown in popularity, and there can be little doubt that the keepers' war on 'vermin' was informed and emboldened by the attitudes of the day.

Although the evidence suggests that many Victorians subscribed to the notion of the terrible killer kite, the view was not universally held. There were some in that era who appreciated the birds' beauty, understood their limitations as hunters and saw the temporary effect parenthood had upon their behaviour. Tragically, these people did little to challenge the killing. Some even participated in it.

In *Short Sketches of the Wild Sports and Natural History of the Highlands*, published in 1849, Charles St John writes, 'That beautiful bird, the kite, is now very rare in this country.' In 1887, Edward Thomas Booth writes of 'these interesting birds' which he fears 'before many years have passed, [will] have entirely disappeared from the British Isles'. Both of these men appeared to mourn the passing of the kite. Both also seemed to understand that they were not the terrible killers they had been cast as. While St John remarked that 'no greater enemy or destructive a foe to young grouse can exist', he balanced this by noting that kites only hunt when driven to provide for their young. Another well-known author from that century, John Colquhoun, stated that 'Buzzards and kites . . . do little mischief to game.' But while it might sound like these men understood the nature of the birds, it didn't prevent them from causing them harm. Each of these men was an egg collector and, in search of samples, killed kites in quite the cruellest of ways.

In *The Moor and the Loch*, Colquhoun describes his attempts to shoot a nesting pair:

> I was perfectly hid and, after waiting nearly an hour, had an opportunity of witnessing the tact and cunning of the bird; at last I saw it flying round in very wide circles, which gradually narrowed: it then lighted upon a distant tree, and peering round in every direction, chose a nearer; and so on, until it came within three or four trees of the nest. It was now within shot.

The bird landed in such a position that a clear shot was impossible.

> Nothing remained but to fire through the nest, which proved a sufficient defence, as the kite flew away, and never returned. A few days after, I climbed the tree with

some difficulty, and took two eggs, about the size of a
hen's, with dusty red spots.

St John (who was responsible for killing the last ospreys in
Sutherland) writes how easily trapped the kite is, explaining to
readers 'one of the most successful ways of trapping the kite is to
peg the entrails of some animal in the shallow part of the water . . .
The garbage catches the sharp eye of the bird, as she soars at a great
height above it, and the clever trapper seldom fails in catching her
in this manner.'

Perhaps the most graphic account came from Booth, the man
said to have killed the last breeding kites in Rothiemurchus. It was
his desire to exhibit taxidermy specimens of every British bird in
all their various stages, from egg to adulthood, in his museum.
(He would ultimately fail in his mission, although he did acquire
more than four hundred specimens. They remain to this day in
the Booth Museum, Brighton.)

Booth went to near unthinkable lengths to add to his collection
– legend has it that he kept a train on steam for a week, ready to
go the moment gamekeepers sent word of rare bird sightings. In
1876, acting on the directions of local gamekeepers, he hid out in
the woods below a kite nest in Speyside, hoping to get a shot at the
adults. His first day brought no success. He returned to the woods
the following morning, bringing with him a bantam cockerel
which he tied up in the nest. Booth continues the story:

> I imagined that when the Kites came in sight and found
> the intruder in possession of their nest, they would
> immediately come within range in order to turn him
> out. No sooner, however, did one of the old birds appear
> than the little cock, who had previously sulked and
> remained laying on its side, raised himself up and, with a

shrill cock-a-doodle-doo, bid defiance to the Kite, which
instantly turned tail and rapidly disappeared from view.

Having removed the bantam, Booth left the woods. He stayed
away some six days, allowing the kites time to settle. Upon his
return he saw one of the birds circling overhead.

> When I fired it fell, as if dead, into the top of a fir, and
> then giving a flap or two of its wings freed itself from
> the branches and went over the precipice into the gorge
> below. When the bird struck the tree I did not fire the
> second barrel, as it was within twenty yards, and would
> have been blown to pieces.

Although he searched for it, the kite was never recovered.

In a final, desperate attempt to catch the remaining adult, he set
a trap in the nest, having first tied the two chicks to prevent either
'specimen' from being snapped in the jaws of the device. When
he later checked the nest, no food had been brought, the trap
remained unsprung. 'We accordingly took the two young ones,
and left the trap which we had previously set.'

The final part of the story is perhaps the hardest to read.

> The nestlings differed considerably in size. One was
> almost full-fledged, while the other showed about half
> its feathers just spouting through the down ... As I
> imagined they would make more interesting specimens
> if slightly more fledged, we kept them alive for a few
> days before sending them to be preserved. They were at
> first remarkably shy and sulky, hiding their heads, and
> evidently disliking to be noticed. For the short time they
> remained under our care, their diet consisted of raw beef;

and on this they appeared to thrive well, though they
resolutely refused to swallow a morsel unless placed in
their mouths and forced half down their throats.

The following year, 1877, Booth achieved his goal, adding an adult
(plus another chick and two eggs) to his collection. In subsequent
years he'd gather more.

Some contemporary writers and conservationists tend to place
significant blame on egg and skin collectors for the decline of
Britain's kite populations. Yet as Bob McGowan, Senior Curator
of Birds at National Museums Scotland (NMS), told me: 'There
is little evidence from museum collections and literature sources
to support this. Only twenty-nine clutches of eggs are recorded
as taken in Scotland between 1860 and 1902, and by 1914 NMS
only held one skin of a Scottish-taken red kite, collected in 1835.
By comparison museums generally have many more clutches and
skins from other rare birds of that era, yet these species were not
extirpated. If, as some claim, skins and eggs of kites were indeed
being collected in numbers which led to population decline, it
seems odd that so little of this material now exists. Given that
context, the argument that collectors *per se* were responsible is
rather weak.'

We should also remember that collectors were often given access
to the nest by gamekeepers who would otherwise have killed the
birds themselves. Make no mistake, it was the massacres carried
out year upon year in the name of game preservation that did the
decisive damage. Next to that, the deeds of Booth, Colquhoun
and St John, horrific though they seem to us now, were a drop in
the ocean.

That is not to dismiss their significance, however, for their
records are vital in other ways. Before samples were gathered
and studied we knew so very little about birds. As high-quality

binoculars weren't readily available, few would even have seen these creatures so close at hand before, far less understood anything of their physical make-up. Even now we continue to learn from the collections. Many are housed in museums and are a first point of call for researchers wishing to know, for example, about the natural history and breeding distributions of birds. By killing animals we learned, and continue to learn, about them. There's an uncomfortable irony to that.

Their records are important in another sense, too. Through their writings we gain an insight into Victorian society. We see a bloody age, where predators were generally reviled and wildlife was routinely killed. In this era, even little boys shot birds with catapults. It was the norm. That some knew what little threat kites posed and still killed them in such violent ways seems to confirm this impression. And if these experts, who worked in the countryside, spent time observing kites at close quarters and cared somewhat for their preservation, could be so brutal towards them, it can be little wonder that kites were finally extirpated. They simply had no one willing to make an unequivocal stand on their behalf, no one to challenge the greedy gled fallacy. If some today still believe the myths, if some would still harm kites, it is surely a legacy of that era, when such thoughts and acts were seen as legitimate.

14

This Is the Life

[The] Kite was believed to nest in a locality in the north
of the area as late as 1877 . . . I remonstrated at the time
for having disclosed the site, but was answered by the
person who recorded the fact that he 'did not think that
there was much chance of outsiders invading the locality,'
and gives us the information that 'keeper' in Latin is
custos . . . Two gentlemen, who are well known as careful
recorders, informed me: 'We were told in 1902 that
the "Gleds" were extinct there now,' the last one having
been killed more than a dozen years ago . . . Alas for the
custody.

J. A. Harvie-Brown, *A Fauna of the Tay Basin &
Strathmore*, 1906

There are moments of sadness in every nesting season. Spring is as
much a time of death as of life. In previous years chicks have blown
from the nest on windy days and been found broken on the ground
below. Keith has climbed trees to discover full-bellied pine martens
on the nest, the young birds devoured. It was no different in 2018.

We found chilled eggs, a nest that was empty but for bits of buzzard chick – 'Martens, Tom,' Keith shouted down to me. 'It's like
an abattoir up here.' A tawny box filled with blood, feathers and
claw-marks. Martens at work again. Nature can be ugly.

The positives outweighed the negatives, however, and the things I saw in my first nesting season will stay with me for life. Peregrines shooting like bullets from a cliff-face, whole sandy banks excavated by badgers, barn owls gliding like ghosts through the broken windows of an abandoned house: this whole world had been on my doorstep for so long and I'd never realised.

Of all the memories, the most thrilling was when Keith climbed up to a nest for the first time. Works of fiction are stocked with moments when time stands still. I've seldom experienced this sensation, but when he opened his mouth to tell me how many, if any, chicks were in there, the feeling hit me. Weeks of hard work were defined in those seconds.

We returned to the woods to ring the kite and buzzard chicks in June. A kite nest is a fascinating thing to see. These birds are terrible kleptomaniacs, stealing all manner of things to adorn their homes. In the days when they frequented Britain's towns and cities they had a habit of stealing caps from the heads of men walking to work. Such thievery continues to this day. From socks and gloves to women's underwear, we've found so many strange items over the years. Once, a severed teddy-bear head was in there. Various theories have been put forward as to why kites behave in this manner. Some think that clothing is intended to strengthen the lining of the nest; others say it is to soften it. In his book *The Red Kite*, Ian Carter suggests that the items are a form of decoration, warning other birds that the nest is in use. All theories, I suppose, are possible, but Carter's makes most sense to me. Were I a rival eyeing the nest, the head of a decapitated teddy would certainly scare me off!

Given the beauty of an adult kite, it is something of a surprise when one sees their chicks. A kite egg is roughly the same size as a hen's and when the chicks hatch they are little more than white balls, all beak and dandelion fluff. To look at this tiny alien lifeform, one would never think that, in two months, it will have transformed into a beautiful, fully grown kite; a day-old kite and an eight-week-old are about as alike as the caterpillar and butterfly.

Three weeks after hatching the transformation from alien to kite is well underway. Downy fluff remains on the head, but immature feathers, known variously as pin or blood feathers, have started to grow. For the first time, it's possible to believe that the bird before you is indeed a kite. We leg-ring them at around the four to five week mark. The parents circle high overhead, waiting for us to leave so they can return. The bond with their young is strong. Once the chicks have hatched, only the most colossal disturbance would make the parents desert them.

In the end, 2018 was a poor nesting season on Argaty. We found only three successful kite nests, ringing just five chicks. (To put this in context, in 2017 Mike had ten nests and nineteen chicks.) I wondered how much harm the frozen February did. In cold weather, food is harder to find; some birds may have been unfit to breed. The heatwave that followed was scarcely any more use. Although food became easier to find, chicks can dehydrate in those long, rainless weeks. Much as I wished to blame the weather for it all, privately I knew there must have been nests I'd failed to find. I'd spent so much time searching for them, walking until my feet were blistered, but on the days that he could join me I realised how easily Keith spotted nests I'd missed. This frustrated me. I wanted to be better.

Next year, I told myself. Next year you will be.

In July, Keith and I took a final trip out, to a barn owl nest I'd found by accident weeks earlier. Barn owls are always susceptible to bad weather. They can tolerate a mild Scottish winter, but a cold one can be fatal. Mike used to have several pairs on the farm, but after successive big chills only one remained.

Last year they'd nested in an abandoned house miles from

anywhere, but when I'd climbed through the broken window and searched the floors earlier that summer I'd seen no fresh pellets. Those I found were dry and brittle. It was a long time since they'd been coughed up. Winter had been so protracted that I had expected this and, so as not to have wasted my time, had brought a kestrel box with me when I set out that morning. If there were no owls, at least I could install that somewhere. Near the ruined house were several old trees. One of them would do nicely. I opted for an ash, which slanted at such an angle that climbing it looked easy. No sooner had I leant against the trunk than, from the far side, two owls soared out and floated silently away, white shadows moving through the sky. Heart racing, I stood watching them, struggling to still my breathing.

Having climbed up and craned my neck to peer around the trunk, I saw that the tree was hollow. A branch had severed, leaving a chasm as deep as a well. I shone a torch down, but the light penetrated only so far. Had this tree stood any nearer to civilisation, someone would have chopped it down. Up here, in this wilderness, it had been left to stand, providing an ideal shelter for the owls.

Ever since that first visit I'd continued to check the tree. Each time the birds were hunkered down deep inside its belly. They flew whenever I came near. I became convinced that this was where they'd nested. Keith brought an endoscope out one day and dangled it like a fishing line, looking for the chicks.

After several minutes' searching he shook his head.

'Cannae see a thing. Hollow's too deep. Probably goes all the way to the ground. No wonder they fly out whenever we touch the trunk. It must echo like a drum. Even if they are nesting in there and not just roosting, we'll never get down to the chicks.'

As he rolled up the endoscope I decided to take a walk around the house. The old shepherd's quarters had been abandoned to the elements decades ago. It looked as if the occupier had just stood up and walked out. From sofas to carpets to tables, most of the furniture was still there. The place was littered with little bones, the remains of a great many voles eaten, then coughed up, by the owls over the years. They broke beneath my feet as I walked the wooden floors.

The house had an outbuilding, an old byre. A black-faced ewe had died there. Her fleshless skull lay half-buried in the mud. Suspended from the rafters, beneath the last remaining part of the roof, was a lidless plastic water drum Mike had strung up a few years earlier. Like many of his recycling projects, it had worked well. In some years the owls had nested in there in preference to the house. On every visit I had checked it just to be sure they weren't inside. It was always empty; the birds were never anywhere but the tree. As I stood below it that day, there came a sound like a gas leak. I put my ear to a crack in the wall, convinced that what I'd heard was the wind whistling through it, but no. The sound was coming from above my head. It was coming from the barrel.

It hung in the corner of the byre above an old sheep pen. I climbed the gates gingerly, afraid that the rotting wood would give way beneath me. To my relief, it held. I steadied myself, turned on the torch on my phone and prepared to peer inside. Some primal fear always kicks in when you look into a nest box. You know that if something is in there it won't hurt you, but perched precariously before it you feel very exposed. Close now, the hissing sounded less like a gas leak, more like a snake. Feeling more vulnerable still, I shone the torch in. Three little barn owl chicks blinked back at me.

Several weeks later, when the birds had grown a bit, we returned to ring them. The sky was clear blue and cottongrass rippled in the field before us. The scene was perfect, a reminder that raptor workers are the most privileged people on earth. As we walked, kites wheeled overhead, scanning the long grass for voles. We passed a kestrel box we'd put up earlier in the year. To my delight, a pair had used it. Five of their chicks had now flown, but one smaller head was still visible, peering out at us. In the next week it would fledge too. I felt proud to see it there in the home we'd provided. Conservation can be pleasing like that. The rewards are tangible.

As we'd hoped, the owlets were now the perfect size for ringing.

Since our last visit, one chick, the runt of the litter, had disappeared. As a precautionary measure, one of the brood is often eaten by its siblings, particularly when other food supplies are scarce. The weight that the bigger chicks gain from this boosts their chances of surviving. As he removed the two remaining chicks from the barrel, Keith, always annoyed by these things, gave them a mock disapproving look.

'What'd you do that for?' he chided, wagging a teacherly finger at them.

One of the owls made us laugh. While his sibling was being ringed, he kept waddling off, moving like a little Napoleon, bum wiggling, wings outstretched, a sour expression on his face, to hide in a clump of rushes. A white bird rather sticks out against an all-green background.

As Keith gathered him in, measured and ringed him, I sat back, absorbing everything I could of this, our final ringing experience for the year. Here I was, sitting mere feet away from two barn owls. What a privilege!

As if he'd tuned into my thoughts, Keith looked up and grinned. 'This is the life, eh, Tom?'

And he was right, this *was* the life. And it was more than that too. This world of blistering sun and pounding rain, of cottongrass fields and tar-black bogs, of wild creatures, piercing claws, gaping beaks and beating hearts, this was life. Life itself. And with its butchered buzzards, stomach-slit frogs and fleshless blackie skulls, this world was death too. And it was brutal and horrific and real in a way that the everyday world, with all of its cosy domesticity, was not and could never be. I wanted this wild world, this connection to something more primal, to something that could never be static but was always changing and evolving, to something that set your heart racing. I needed it.

We made our exit obvious, walking out in the open so that the mother owl knew it was safe to return to her young. On our way back to the car we said little. When Keith had returned the birds to the barrel our moods had quickly changed. An air of sad resignation now hung over us. It was all over: every chick that we could find had been ringed. When we first began our monitoring, Keith had warned me about this point. I knew to expect this empty feeling. (He'd called it depression; now I could see why.) Before long we'd begin this work again, clearing the mess out of nest boxes, replacing any broken ones. In a matter of weeks the kite chicks would appear for food at the hide and that would be wonderful, but for now there was nothing to do and that emptiness was awful.

I knew it would always be this way, no matter how many seasons I completed in the future. I knew it. Because this was the life.

15

The Missing Pieces

A second cleugh in another of the glens was known as
Gladscleugh and was possibly an ancient haunt of the
glead or kite. The last one in the district was trapped in a
pole trap not far from Gladscleugh. Was he visiting, like
the eagle, under the pressure of some hereditary impulse,
the home of his ancestors? The bird may be seen still, a
beautiful specimen of Milvus regalis, in a case kept in a
private home in the town.

The Scotsman, 9 April 1912

In late summer the young kites fledged. They circled over the
woods they'd grown up in like children running around their back
gardens. I was thrilled to see that they'd made it, to watch them
flying free. A parent, probably the father, hovered protectively
close. For the next few weeks he'd feed them, teach them to hunt
and scavenge, give them what tools he could to survive in the outside world.

The squirrel kittens left their dreys too and it didn't take them
long to find our food. I'd expected them to be much smaller than
the adults, but the difference was not so great. Their tails were
thinner, more rat-like, but otherwise they looked similar to the
rest. What truly marked them out was their confusion about the
feeder boxes. They could see and smell the food inside, but had

no idea how to get at it. Some tried to chew through the wood, others pounded the Perspex front, demanding admission. While engaged in these assaults they often butted the lid open by mistake. It swung back down, clouting them on the head. In this manner, learning from the most painful of mistakes, they discovered how to get in.

At the same time that the young squirrels arrived, masses of fluffy juvenile tits – blue, great and coal – also joined the party. Most would remain thereafter, drawn in by the steady supply of food.

With the hide now open and photographers hiring it regularly, my time there was more restricted than before, but even a quick trip down to fill the feeders had a restorative effect. More and more my mental health seemed to depend upon the squirrels and the birds, the beautiful things and the wild worlds. The fragility of this happiness was apparent too. In Dumfriesshire a squirrel pox outbreak had recently been reported and a harrowing photograph had found its way on to the internet. The infected red squirrel looked normal in all respects bar one. Where its eye should have been a livid pink welt had erupted. The poor creature had dwindled away and died a few days later. I couldn't shake that image from my mind. I lived in dread of grey squirrels arriving here, of the devastation that the virus would wreak.

In July, Radio Scotland's *Out of Doors* programme visited us again and I spent an enjoyable hour with presenter Mark Stephen watching the squirrels and birds, chatting about wildlife. Mark was an easy person to talk to. Open, thoughtful, the type you trust instantly.

'This is a lovely thing you're doing here,' he said later, when we'd finished the interview and were heading for the car. Prompted by that, a theme that had been circling my brain for months found its way out into the light.

'We're lucky to have all of this,' I replied, taking in our surroundings with a glance. 'Most people don't have this opportunity and the few that do don't often use it for any good purpose.'

'You're right,' he said, nodding. 'Most have a house and garden . . . if they're lucky. At best they can help a few wild creatures. With a garden this size, you could help thousands.'

That gave me pause for thought. Ringing season had demanded so much of me that I'd had little time and energy to consider anything as large as my aspirations for Argaty. That goal, set out in April during dinner with my parents and our friends, remained roughly the same. Back then I'd spoken of turning this place into a world-class wildlife watching venue. Revisiting the mission statement now, I saw that it was close to what I wanted, but not quite there. World-class. Venue. These words were the problem. Too much they reflected my old mindset rather than my current one. Setting off down this road a year earlier, I'd been driven by the need to show myself, and anyone watching, that I could do this. I was performing to a crowd. And yes, we still wanted and needed an audience for the work we were doing – without them we would have no income and the business would fail. But the audience itself was not the motivation for continuing. When I boiled all of this down to its essence, and everything less solid evaporated, just two questions remained: what really mattered to me, and why? The important thing was the animals, the squirrels, the kites and the wild things that I loved. Making a home for them and anything else I could was the real goal. Why do this? Because we were lucky, and many others weren't. Because we had a chance that others didn't. Because we could make this place a home for wildlife. And if people like us didn't use our opportunity, then wildlife wouldn't be saved.

To make our home a home for wildlife. That was our goal.

Because we could. That was our reason.

And should any of this seem pious, false, too earnest to be credible, I should also make this confession: it probably is. I know that we have more than others, and I do feel bad about that. Doing something positive with my privilege is my best means of quelling the guilt.

So we plunged into new projects, enlisted new volunteers and established partnerships with experts who would show us what to do to achieve this aim.

Mike's bumblebee garden had rather gone to ruin in the past decade. As a final collaboration in the weeks before he left, we had set about revitalising it. I continued that now, aided by our amazing new team of volunteers. With the sun toasting our backs and kites passing overhead, we waged war on the tall grass, pruned back trees to allow light in, planted new nectar-rich flowers where the old ones had been crowded out. We built a deck beside the pond in the middle of the garden and bought nets so that children could dip the water. On the days when my mum looked after her, Rowan would often wander along the farmyard to inspect our progress. We shared many lovely hours dredging the water, looking for frogs and newts. I love to think of her sitting there, little wellingtoned legs dangling off the deck, eyes contracted in concentration as she peered into the net, inspecting our catch. The memory always makes me smile.

In those sultry summer days the bees came out in abundance, drawn in by the new plants, and at night so did the moths. In a moth-trapping event we caught several hundred. What a glorious thing to see the elephant hawk moth, the tiger, brimstone, smoky wainscot and spectacle moths (so named for raised tufts of scales on its thorax, which resemble a pair of glasses) up close.

Ruary Mackenzie Dodds and Kari de Koenigswarter ran an Introduction to Dragonflies course at the centre. More than perhaps anyone else, Ruary and Kari had put these wonderful insects on the map in Britain, establishing the nation's first-ever dragonfly sanctuary and inspiring people to care about their fate.

Their enthusiasm was infectious. I could see why they'd converted so many people to the cause.

'Lovely question,' Ruary would say, beaming, whenever someone asked him anything. Although he must have fielded the same queries millions of times, he greeted each one as if he'd never been asked it before.

He and Kari taught us the differences between dragonflies and damselflies (dragonflies are bigger, they live longer, have eyes in the centre of their heads and keep their wings outstretched like an aeroplane when they land; damselflies' eyes are on the side of their heads, and almost all of their kind tuck their wings back when they land). We learned about their life cycles and how to identify different species. Later we visited one of Argaty's twelve ponds. That number had never seemed particularly remarkable to me before, but when we told Ruary he smiled again with delight.

'Wonderful,' he said.

The smile grew larger when we arrived at the water's edge. Everywhere you looked there were dragonflies. They were climbing the rushes, skimming the water, sunbathing on floating leaves. Ruary jumped around like an excited schoolboy, pointing out flying jewels in the distance, while Kari knelt quietly by the water, scooping out larvae to show the children, pointing to the rushes where dragonflies had shed their skin. What a lovely team they made. I watched them carefully, considered the ways we might help in the future. Promoting dragonflies seemed both a worthwhile and an exciting cause.

The hot weather continued. Ahead of schedule we brought the hay in, took first then second cuts of silage. Kites followed the tractor, buzzards perched on the freshly rolled bales, their keen eyes searching for invertebrates and small rodents killed when the grass was chopped. When we cut the field, we always begin in the middle and work out the way, allowing as many of the residents as possible to escape to the margins. (Cutting from the outside in would leave all the wildlife – rodents, hares, rabbits – trapped on an island in the middle. Mowing that last patch would inevitably lead to carnage.) Some things don't escape, no matter how hard we try. The raptors fare well on these days.

While spiking silage bales with the tractor forks, preparing to take them away, wrap them in black plastic and stack them for the winter, my gaze repeatedly fell upon a tall tree in the nearby woods. Several years earlier Mike and Keith had built an osprey platform in the top of it, one of three they installed on the estate. Every summer since, an osprey had visited this one. He left Africa in spring, headed over Europe, crossed the Channel and sped north through England. His goal was always to get here, to this tree, to this nest. Each year he added more sticks to it, showing that he could build, could make a home fit for a family. Then he did laps of the local area calling out for a mate, asking her to share his life. He'd never yet found one.

From the tractor cab I watched him return periodically throughout the day. As usual, he came home alone.

*

In August, I took some friends to Argyll on Scotland's west coast. Earlier in the year the photographer and naturalist Philip Price had visited Argaty and delivered an inspiring talk on beavers, whetting my appetite to see them. Once widespread across our countryside, beavers were hunted for their meat, their pelts and for castoreum, an exudate from their castor sacs which was used in medicines and perfumes. By the sixteenth century most are thought to have been wiped out in Britain (though isolated records exist from as late as 1789).

How our landscape must have changed in the beaver's absence. Had we not removed them, what might our rivers look like now, how many more wetlands might we have? One can only speculate. We can say one thing with certainty: had these masters of habitat creation been here all along, the UK would not be languishing in 189th place in the State of Nature rankings.

The beaver's tale began to change in the early 2000s. Farmers began reporting sightings of them in Tayside. It transpired that animals from private enclosures had escaped into the River Earn. A government-sanctioned trial release began at Knapdale in Argyll in 2009. The purpose of the trial was to establish what effect beavers could have on the local economy, whether reintroductions to other parts of the country might one day be viable, and what impact they might have on the surrounding landscape. Of these questions, it was the third that most interested me. We journeyed to Knapdale in search of an answer.

Philip's car stood out immediately from the others parked outside the Gallery of Lorne Inn. Sat like Napoleon's bicorne hat atop its roof was a huge three-person canoe.

'We're going to my favourite beaver-watching spot,' he told us. 'Hardly anyone knows they're there. We can take a paddle out and watch them in total peace.'

As evening drew in we followed his car along a deserted Forestry Commission road. At the end of this dark, silent and largely unremarkable trail was a beaver territory. The sound that greeted my ears upon exiting the vehicle was unlike any I've ever heard. In the distance a thousand chainsaws were buzzing, or so it seemed.

'So many insects,' I said, stunned.

'All thanks to the beavers,' Philip replied, loosening the straps that held the canoe to the car. 'They fell the trees, the wood starts to rot and that makes great habitat and food for insects.'

To our right a narrow burn trickled lazily by. Upstream, a tangled blockade of sticks stretched across its width, sieving the water, slowing its flow. Beyond was a lochan the size of a football pitch. Trees formed an amphitheatre around the wetland. Many were down, lying like fallen soldiers at the end of a battle. I'd expected that. Stranger, though, to see others still standing out there in the water. This was something utterly foreign, an unknown world.

Presently there came a plop, like a small stone breaking the water. At the lochan's edge a willow was slumped on its side. Beside it rippling rings were spreading like echoes across the surface. There, just moments before, a fish had jumped.

'When the beavers drop trees, it brings insects closer to water,' said Philip. 'The fish have never had so many easy meals. They're thriving now. Everything's gone up in number since the beavers came here.'

As darkness deepened, bats appeared, taking their turn on the midges (which in turn were nibbling us). The omnipresent drone of insects continued. Then, from the far end of the lochan, came a sound that stopped my breath. A beaver had taken the plunge.

'Time to go,' said Philip.

We set off, paddling gently towards her, drawing so close that, had I wished, I could easily have reached out and touched her. Instead we idled in the water, letting her mosey on past. She didn't even know we were there. Like all of her kind, she was blind as a bat. (For reasons of accuracy, it should be noted that this is simply a figure of speech. Neither beavers nor bats are in fact blind!)

Seeing that landscape, how the beavers had changed it, how they'd created life, left me very emotional. It challenged everything I thought I knew about what the countryside was or could be. Only in the presence of a beaver did I realise what an environmentally depleted age I have lived in. This land, where trees stood in the middle of water and life functioned independent of people, was so alien to me, and it shouldn't have been. This was what so much of Scotland was supposed to look like. To witness it was to see a country that had been lost, that our forebears stole from us when they killed the last of the beavers.

I've heard it said that were it not for the success of sea eagles and kites the beaver project (the first reintroduction of an extirpated mammal in UK history) would never have happened. To a sceptical public, birds are an easier sell. They may impact their surroundings; beavers transform them. But thanks to these birds, we now know that we can bring things back, that they can survive here and that they will generate money.

How satisfying to see kites as part of this picture, a stepping stone to something bigger. Because we brought the birds back, we were able to bring the beaver back. If we manage to bring beavers back, might we one day see animals like lynx roaming these lands again?

We need them here, of that there can be no doubt. Contrary to popular opinion, the absence of top predators does not ensure that the rest of the food chain will thrive. Quite the opposite is true. For a time, populations may increase, but then food becomes scarce, diseases become prevalent. A healthy ecosystem needs killers. The estimated 400,000 red deer in Scotland provide the classic example of this. (The Deer Commission for Scotland, the former government body responsible for the conservation and control of deer, suggested that we have enough habitat to sustain 60,000.) My hope is that, much as the raptor reintroductions made the beaver's return possible, the beavers will open the door to other missing mammals.

Although there are very few records of wolves ever hurting people, 'Big Bad Wolf' stories have a long legacy. Given the public fear of them, a reintroduction would seem very unlikely in the near future, no matter the ecological imperative of bringing them back. Someday, hopefully, it will happen, but Britain has been steering this course for a long, long time. Our ship will be slow to turn. Lynx, however, should be highly possible. Masters of surprise, these medium-sized cats lie in wait, then burst out from cover to ambush large prey. What a difference such skills could make here. Across their European range, deer make up a high-proportion of a lynx's diet. Our deer populations are so great that we will have to cull them for years, maybe generations, but with lynx to help us, our ineffectiveness as a hunter would matter less and less. With fewer hungry mouths browsing upon saplings, nipping any chance of regeneration off at the stem, we might at last increase our pathetic levels of woodland cover.

A reintroduction might cause some difficulties. Farmers in particular worry that sheep may be predated. In truth the shy, secretive lynx tends only to attack sheep that are grazed in woodland, as happens in Norway. The UK doesn't have much woodland and we tend not to keep livestock in it anyway. Having an animal back that suppresses fox (and rabbit) numbers would, one supposes, bring a net gain for farmers.

But I have, I confess, great sympathy for farmers when it comes to reintroductions. Although they are a sizable minority, farmers are nevertheless hugely outnumbered in our country and I'm quite convinced that they feel it. Producing food in this climate is hard enough; the thought of bringing back animals that could make the job even harder is understandably unpopular.

The trouble is that farmland makes up three-quarters of Scotland's land mass. Quite simply, there is nowhere else for us to reintroduce animals. This will impact people's lives, but we need these missing predators back if we are serious about balancing our food chain and restoring natural processes. While I wish this weren't being forced upon farmers, the climate crisis and wildlife crash are bigger than any other concerns. The science tells us, irrefutably, that we must act now or millions of people will die, animals and plants will go extinct, the world as we know it will collapse.

All of this was running through my mind that night in August as I watched the beaver swim leisurely across the lochan. In that scene I saw both what we have lost and what we might have again if we embrace reintroductions. Britain's natural jigsaw has been broken up. Its pieces lie scattered. Until we find them and put them back in place our ecosystem will never properly function, our picture will never hang together, we will contribute little to the fight against mass extinction. We need these missing animals back.

16

A Pleasant Blur

That the red kite still survives in Central Wales, but nowhere else in the British Isles, was confirmed in a detailed report on the present strength of the bird submitted to a special meeting at Carmarthen on Saturday.

Western Mail & South West Mail, 20 November 1950

Summer's fire was fast burning out. The trees began to moult, then the rain finally came. On those sullen grey mornings the kites flew low, peering through the murk, searching for food below. On my walk to work I often saw them land in the field and pilfer a worm as it left its burrow. How ungainly the kite looks on land, lumbering around, tail dragging through the mud. This bird belongs in the sky.

Throughout the year our visitor numbers had slowly grown. Our social media posts were gaining more attention too. This was in large part thanks to our events and evening talks, which were proving more popular. Trying to attract high-calibre speakers or event leaders was not always easy. Coming into this, I'd known no one in the nature community. To climb up from the very bottom I'd had to make some friends. Striking the balance between forming relationships and networking is tricky, though. When does asking for help slip into social climbing? Those were very real concerns, but what option was there but to make unsolicited calls in the hope that I'd seem genuine to the people I was approaching? I'd thrown in my lot with the kites and progressing the business. I had to make it work.

Fully expecting a chorus of rejections I took a chance and contacted three of Scotland's greatest wildlife writers. To my amazement they all accepted invitations to give presentations at Argaty. Polly Pullar delivered a beautiful, meditative talk on Ardnamurchan, where she had spent much of her childhood and where, more recently, she had returned to write a book on pine martens and the other varied wildlife that inhabits that great, untamed peninsula. When reading her books I am always struck by her empathy for nature. It courses through her pages. Because she cares, she makes you care. That's her greatest gift. She would become a great friend, supporting me and promoting the project at every opportunity.

Author and retired wildlife detective Alan Stewart was next to visit. He answered many of the questions I had on the difficulties of preventing wildlife crime in Scotland. As most of the incidents occur in remote places it is hard to catch the criminals, harder still on the large estates to prove which of the many people working there commit the crimes. Even with the required evidence, several of Alan's cases were time-barred after the procurator fiscal failed to deal with them in the required period. And of those that did result in a conviction, the punishment seldom fitted the crime. Often a paltry fine was all the criminal received. In Scotland we have some of Europe's most advanced wildlife protection laws but little means of enforcing them. No wonder crimes still happen.

Our final guest, Jim Crumley, stood up without a slide show or notes for backup and simply talked and read to us. I loved the immediacy of that, loved hearing those words being delivered by the person who wrote them. When I read I skip words that I shouldn't, stress the wrong ones, miss the overall picture. Hearing Jim, understanding how the words build the sentences, the sentences the paragraphs, I realised what an immense act of construction our best authors do. Every word Jim Crumley writes has a purpose; not one is wasted.

Looking back, it amazes me that I could simply call up these wonderful writers and they would agree to come to little, unknown Argaty. How could it be so simple? In any other sphere I doubt it would be, but the wildlife world is different. This is a small group of compassionate people united by a common love. True, I wasn't a known face, but that didn't seem to matter. Our work with the kites gave us all the credentials we needed. We wanted to grow the business and our new friends seemed happy to help. In return we'd help promote their cause. This was good old-fashioned bartering at its very best.

*

The days grew shorter, the weather grew worse. Holed up in my office on those dark nights I began reading my great-grandfather Charlie Bowser's journals, learning more of the history of Argaty and of my family. The diaries cover his childhood in Glasgow and his early period as laird of Argaty in the 1920s and '30s. After the death of his father-in-law, Charlie took over the management of the estate. Tenanted farms that had been poorly run were taken back in hand, new agricultural methods were employed, new sheds built. Houses were constructed, too, and woodlands planted. All of this is chronicled.

I never met Charlie Bowser. He died before I was born. But he wrote with honesty, compassion and a self-effacing humour that laid him bare to the reader.

The first entry in the journal was a sign of things to come:

> I enjoyed my schooldays to the full without, it must be
> confessed, any more concern as to what it was all about.

Somehow or other I got away with it . . . Many years later,
on meeting one of my schoolmasters, I was comforted
in his recollection – You were not clever, but you were
highly intelligent.

The more I read, the more I found to like about my great-
grandfather. There were hand-drawn maps of woodland forts
and bases that he built with his children, loving letters written
to his wife, an evident concern for the natural environment.
He seemed like a good boss too, a man who cared about the
people on his estate. This excerpt, written in 1928, I found very
touching:

In the middle of September James Drummond Head
Forester died. He was an efficient man at his job and
a most lovable man full of stories and bound up in
Argaty as much as anyone thereon. There was nothing
he saw outside of Argaty that could not in his opinion
be equalled or surpassed at Argaty. The plantation at
Lerrocks which was planted in his lifetime and planted
after his death is to be called Drummond's Brae . . . a
house has been bought in Doune to accommodate
Mrs Drummond and her family.

Reading those words, I could imagine my own father saying them.
The language was similar, the tone virtually identical. I thought
of the people and environment around me, saw the ways in which
they had come to be as they are. Both were a product of the people
who came before us. The more I read of the diaries, the more I
understood that.

The remainder of 2018 passed in a pleasant blur. Never in my
working life had I been so busy, nor so content. In September

Argaty featured on *Landward*, the BBC's farming and countryside television show. The wind howled all morning and the rain never ceased. Unable to hear predators, the squirrels laid low for hours. We sat in frozen silence, all conversation long since exhausted. From time to time the wind died down and I said a silent prayer, begging it to stop entirely, but each time it came back with greater anger than before. The rain drummed on the roof, the trees shook, the squirrels stayed away.

'I'm so sorry,' I said for the umpteenth time. 'This bloody weather.'

At last, when I'd long since abandoned hope, hunger got the better of the squirrels and they ventured out. I could have wept with relief. By the afternoon, conditions had improved and the kites at least performed admirably, sparing my frayed nerves from further torture.

When the show aired, my nearest and dearest mocked me mercilessly for my five minutes of fame. I don't usually speak with much of an accent, but when the camera was rolling something came over me. Unconsciously I adopted a faux-Scots television voice, enunciating every syllable as though the presenter, Arlene Stuart, was deaf.

Rowan gave me the sort of horrified look that four-year-old girls reserve for embarrassing fathers, then burst into fits of laughter.

'Squirrrrrel,' she mocked, rolling her Rs like a purring cat. 'Rrrrred kite.'

Worse was to come from my dear wife and our friends. Towards the end of the feature, I grew a little too comfortable with the idea of Tom the television personality.

Speaking to camera, Arlene said, 'With wildlife tourism on the up, Tom is obviously quite keen to diversify and move forward. Only one condition, though . . .'

'Yeah,' I said, laughing at my own ingenuity, 'so long as the animal's red.'

That joke would haunt me for months.

So long as it's red. Idiot!

Sarah and my friends repeated these words so often I still hear them in my sleep.

As it happened, I might have spared myself all of this abuse, for the next diversification I had in mind wasn't even red. Months earlier I'd begun feeding squirrels in other woods around the farm. This extra feeding wasn't part of any business plan; I didn't want to build additional squirrel hides or make money from this feeding. Little acts like this simply fitted the wider ethos of using our land for some good.

One morning I noticed that one feeder box was hanging at an angle, as though something rather heavier than a squirrel had set foot upon it. The nuts inside, which usually lasted days, had gone. A stealth camera would prove what I already suspected: a pine marten was working the area. Once she had discovered the food, she returned again and again. Eggs were very much her favourite delicacy. The camera recorded her sneaking in, grabbing them between her teeth and slinking away like a frightened cat, her belly grazing the ground. I'd find the shattered shells in the bushes a few yards away.

Earlier that year I'd had and lost another marten. For a short spell, it had appeared on camera every night at the woodland hide. The prospect of finally seeing one in the flesh had me so excited that I'd rushed things and set a night light beneath the feeders so the whole area would be illuminated; the spooked marten never came back. This time, agonising though the wait would be, I was determined to do things correctly, feed her through the winter in the hope that she'd have young and form a real attachment to the spot. Only then would I try lighting the area. We had another marten, another chance. At night, when I drifted off to sleep, the mysterious, nocturnal killers described in the old naturalists' accounts stalked my dreams every bit as much as they stalked the woodlands of Argaty.

November arrived and I continued my nightly rendezvous with my great-grandfather's journals. A whole world was there, preserved forever on the typewritten pages of the hardbacked book. This was a world of serving staff, hunting and fishing, of holidays to the Continent while nannies took the children on freezing but character-building seaside jaunts to Ayr or North Berwick. Charlie travelled to Twickenham to watch Scotland play England at rugby. Charlie bought horses and romped them around the estate. Charlie commissioned Philip de László to paint a portrait of his wife. Charlie really lived.

A sense that I understood my great-grandfather grew with each page. He was such a great recorder of life: my great-uncle Hubert casting his baby teeth, Hubert and David (my grandfather) sledging for the first time on the lawns outside the house, Eleanor Hope (my great-aunt) taking her first infant steps. Not only did Charlie write, he also made films. Every Christmas he gathered family, friends and estate staff and played them footage on cine film reels he'd shot that year. Now, nearly a century later, here I was thinking about the same things he did: family, Argaty and the wider world. Later, when I began work upon this book, I'd write about them, too. That shared purpose, that desire to record, tied me to my great-grandfather.

The year moved on and I continued to read.

In the closing days of November, we heard some exciting news. The shortlists for the RSPB's Nature of Scotland Awards were announced and we were in the running for two categories. Although I'd spent hours on the applications, writing down every good turn we'd ever done for Mother Nature, I hadn't expected to make it that far. My parents and I attended the black tie dinner in Edinburgh and received another surprise – we won the Nature Tourism award.

The *Landward* feature played on the big screens, hundreds of people died inside at the 'so long as the animal's red' joke, and we took to the stage to accept the award from Kate Humble and Euan McIlwraith.

'I hope you know that your dad and I are very proud of you,' Mum said, pulling me in for a hug as we returned to our seats.

'I know,' I said.

I waited a moment, watched as her amusement turned to incredulity, then smiled. 'I'm proud of you, too,' I said.

When they established this project, my parents took on commitments that many wouldn't have. They swore an oath to the kites, pledged to be here every single day to feed them, to always be their advocates, to always fight to convince the sceptics of their virtues. And though they were supposed to be stepping back, easing gradually into retirement, both had worked harder than ever to keep the kites and the estate going that past year.

Many others have contributed to our project. This award was a recognition of Sarah, who had been better to me than I deserved during this whole period. How impossible I must have been as I lost my old identity completely and tried to forge a new one. Such processes are never painless and I dread to think where I would be had she not stood by me. The award was also a recognition of the RSPB's red kite team, of Mike and of our wonderful volunteers who help us more than they could possibly know. It was a recognition of my grandparents and great-grandparents, who'd managed the place so well in their day. But more than anyone else, the award was for my parents, for the oath that they swore so many years ago. They deserved it.

What Next?

17

Winning?

The sea eagle is ... back with us, patrolling the skies of
parts of our western seaboard and furthermore, successfully
breeding in secret and well-guarded locations. Its presence
here is entirely due to the helping hand of man ... The
red kite, it seems, is to be next. A similar programme of
introduction – again from Scandinavian stock – has begun
and, hopefully, in a few years' time, we will be treated to the
sight of the fork-tailed predators, presently only hanging on
naturally in parts of Wales, in the skies over Scotland.

'Keith Graham's Country View', *Perthshire Advertiser*,
10 October 1989

There's a story I like to tell when on duty in the hide, of something
that happened back in 2016. At the time it seemed fairly insignif-
icant; now I see it differently.

It was summertime and Sarah, Rowan and I had travelled to
Inverness. Sarah's mum, Beth, lives at the northern edge of the
city, just across the water from the Black Isle, the site where the
first kites were reintroduced back in 1989. The house looks out
onto a grassy hilltop skirted by lovely mixed woodland. While the
rest of us sat drinking coffee and swapping news, Rowan, just two
years old at the time, was looking out of the window, watching the
world go by. Suddenly her little finger pointed to the sky.

'Red kite!' she shouted.

'I'm sure it's just a seagull,' I said, without even turning to look.

The Black Isle kite population is one of the few that hasn't significantly increased and I thought it far more likely that she had spotted a gull flying over.

'Kite,' she said again, insistent, verging on tearful, and this time I did turn.

And she was right, of course. There, floating slowly over us, was a Highland cousin of our own Central Scotland birds.

It took me twelve years to learn what a red kite was. Rowan just knew. That little story means a lot to me. It's a sign that our world is changing, that we are making progress.

The question of progress was seldom far from my mind as 2019 dawned. This was to be a landmark year for kite conservation. Come summer it would be thirty years since the birds were reintroduced to the Black Isle and England's Chiltern Hills. Three decades in, how did we assess the progress of the reintroduction projects? How successful had they been?

Mike and I discussed this on the telephone one night. We sat chatting for a long time, laughing, joking, catching up. Things had settled into a pleasant rhythm between us again. Although I still thought of him constantly, still remembered how hard it was when he left, those times seemed so distant it was as if someone else lived through them. The lives we were both now leading felt like the only ones we'd ever led. Time, as they say, is a great healer.

The one lingering regret I still nurture is that I'll never know what we might have achieved together had I only taken this path a few years sooner. Of course, I'd never have done it. I know that. He did too. I was the sort of person who needed a kick up the backside before I was moved to action. Always reactive, never proactive, back then I only ever tried to fix something when it was already broken. And given the extent of Mike's knowledge and the limits of mine, I'd always have felt like the junior partner, the little brother. I'd never have found the courage to take the lead had it not been thrust upon me. Because of who I was then, because of these failings, we could never have taken this journey together. I could only travel alone.

Does any of this matter now? Probably not. Mike is happier and so am I. Kites have been spotted around Brockholes, one of the wildlife reserves he now works on in Lancashire. Though still a rare sight, they're spreading from their release sites in Yorkshire, Cumbria and Gateshead. In time they'll fill his skies again, as they once did when he lived at Argaty. A cheering thought.

During our conversation he talked of his time here, told me things I didn't know. There was the story of the kite who injured his foot so badly he couldn't grab food on the wing. Instead he had to land in order to get anything. That bird's chicks took their father's behaviour as normal and copied him when they fledged. One day Mike saw the whole family sitting in the grass, tucking into the afternoon feed while all of the other kites plunged around them. Was the standard practice of feeding on the wing, which everyone believed to be instinctive, in fact learned behaviour? Then there was the tale of the Inverness-shire kite that was found in 2008 in Angus. It had eaten bait poisoned with the banned rodenticide alpha-chloralose and would have died of hypothermia (one of the poison's infamous side effects) had it not been a heatwave summer. Thankfully, the SSPCA were able to rehabilitate it and it was released on Argaty later that month.

Finally, Mike recalled his first visit to the farm. It was 2003 and he'd just accepted the head ranger job. The builders were still at work constructing our hide, the path to it had yet to be installed, the rowans, hazels and hawthorns that now encircle it were yet to be planted. But although so little was in place, fifteen kites were circling overhead and he knew that it would come together, that the project would be a good one. That image cheers me, too.

I told him of the questions that I'd been pondering, asked him how he gauged the progress of the kite reintroduction projects.

'Things are so different now,' he said. 'At the start there were so many problems with persecution and old attitudes. And yeah, a lot of the issues are still there, but nothing like they were. Kites are on the up. I think we're winning the battle.'

I mulled this over for days afterwards. Was he right? Are we winning? Sometimes it's hard to know. The depressing truth is that kites are still killed every year in Britain. There was a spate of incidents in 2018, some of which have stuck with me ever since.

Four kites were found dead between January and April in south-west Scotland. Tests later confirmed that at least three of them had been poisoned. In the North Pennines a kite had to be euthanised after it was shot and blinded. Another in Wales was found in a gin trap (an illegal spring-loaded trap with serrated jaws). Both the kite's legs were broken and the vets had no choice but to put it down. A breeding pair died of poisoning in Northern Ireland. The female was discovered on the nest, lying slumped over a clutch of eggs. The male was found in distress and died shortly after.

More cases followed, all of them fatal.

Perhaps the one that shocked me most had occurred in May. Two walkers in Yorkshire found a kite lying incapacitated on the ground. X-rays revealed a shotgun pellet in its wing. The bird was euthanised after vets decided it would never fly again. Though this incident was no more disturbing than any of the others I've listed, and arguably less horrific than some of them, it sickened me most of all. Why? Because the bird came down on Harewood estate, where kites were released in 1999 when the RSPB began their Yorkshire reintroduction. The estate owners and their staff have acted as the birds' ambassadors in the area ever since. This kite was believed to have been shot very close to the estate boundary. When I read of it, I tried to imagine how they felt, suspecting that someone nearby had committed such a crime. How might I feel if something like that happened here? And then I remembered the kite that was shot so close to Argaty in 2009.

Although nine years divided them, the cases were so similar. To think of them, it was tempting to wonder whether anything had changed in that time. Has anything, in fact, changed since kites were first brought back three decades ago?

Since reintroductions began, 327 kites are known to have been killed in Britain (figures correct up until end of September 2019). In Scotland three of that total were found dead in illegal traps, eight were shot and a massive 104 were poisoned. Forty-seven of the poisonings occurred in the Highlands. The most infamous of these took place in 2014, when sixteen kites and six buzzards were found dead in the Conon Bridge area. Tests later confirmed that at least fourteen of the kites had been poisoned (the other two bodies had been scavenged and could not be analysed). Although the Black Isle was the first Scottish reintroduction, their population lags a long way behind Central Scotland and Dumfriesshire, and many believe that kites still come to harm when they venture onto driven-grouse moors.

In our area a lot of gamekeepers have rather taken kites to heart. Much of the meat we feed the birds comes from local keepers. As Central Scotland has little moorland, there is little raptor–grouse conflict, but on two of the only driven-grouse shooting estates kites have been killed. In the Gargunnock Hills, west of Stirling, one died after ingesting a banned poison in 2003. Another died of poisoning in the area in 2014, and in 2015 one was found in an illegal trap and had to be euthanised due to the extent of its injuries. A peregrine was also poisoned there in 2015. Although there was insufficient evidence to charge anyone from Burnfoot, the estate in question, the incidents above provided a sufficient case for a General Licence Restriction, which SNH issued in 2016 for 'issues associated with poisoning birds of prey and the illegal use of traps'. The sanctions prevented the estate from any form of avian 'pest' control for three years.

Edradynate, another grouse moor, near Aberfeldy, was also served with a restriction. Alan Stewart, the author and former detective who spoke at Argaty in 2018, described Edradynate as the worst in Scotland for wildlife crime. Between 1993 and 2001, poisoned animals including buzzards, crows, tawny owls, a gull, a pole cat and a kite were discovered in and around the grounds. The further discovery of two more poisoned buzzards in 2015 proved to be the final straw and the restriction was served.

In recent years, stricter legislation has led to a slight decrease in poisoning and trapping incidents in Scotland. The 2003 Land Reform Act allows everyone responsible access to the countryside. With hillwalkers and ramblers on the land, there is every chance that poisoned baits and illegal traps might be discovered. Few estates want to suffer the same sanctions imposed on Burnfoot and Edradynate. But these incidents have not ceased entirely and, despite the available habitat, it's noticeable that there are certain places in Scotland that kites are mysteriously absent from. Our population is growing, but it ought to be larger; these birds ought to be spreading more.

If you are anything like me, you're thinking of the SNH eagle report, of all of those birds whose satellite tags suddenly stopped working; you're thinking of the claim made within the report, that this was both suspicious and indicative of external human influences, for no other explanation could be found; you're thinking, does the same thing happen to untagged kites?

And if you really are like me, you want an answer to the following question: Why? In the twenty-first century, why do these age-old problems still exist? Why would people kill kites?

A few weeks after my chat with Mike, I met Ian Thomson, RSPB Scotland's head of investigations. He and his team have looked into all manner of persecution incidents, including many kite deaths. We sat in a meeting room in the RSPB's Edinburgh office. From time to time, my gaze drifted out of the windows to the Pentland Hills, where a satellite-tagged golden eagle, Fred, vanished in January 2018.

When I put my question to Ian, he leant back in his chair, looked to the ceiling and thought for a moment.

'Why would people kill kites? Doesn't make sense, does it? Could be all sorts of reasons. Some people don't want birdwatchers or raptor workers coming snooping around, particularly if they're killing other protected animals. Even if they've got nothing much against the kites, the best means of keeping people out is to have no raptors there. Sometimes they see a nest up in a tree and just shoot it out. Doesn't matter to them if it's a kite, buzzard or raven up there. And some people simply don't want anything with a hooked beak and talons on their land.'

'It always strikes me as strange,' I said, musing aloud. 'I mean, kites might take chicks but not an adult game bird. But I guess for some estates any kind of loss is unacceptable, especially if they want maximum grouse numbers for driven shoots.'

He nodded.

'Can they do it?' I asked. 'Can they have those numbers without killing everything that predates on them?'

'No,' he said simply. 'Not if they want those amounts.'

Our conversation turned to grouse-moor licensing. In a matter of months the Grouse Moor Management Group (established to examine the environmental impact of driven grouse shooting and to consider the merits of a licensing system) would submit recommendations to the Scottish Government. The same questions were on everyone's lips. What would the review say? Would they recommend stricter legislation? And how would the politicians respond if they did?

I told Ian about the wildlife crime talk we'd been forced to hold in secret a year earlier after the protestors had threatened to picket. Since that time representatives of the shooting bodies, as well as the Scottish Gamekeepers Association and Scottish Land and Estates (an organisation representing rural landowners and businesses), had repeatedly denied the need for further legislation. They quoted one statistic (that records of poisoning in Scotland are at an all-time low) and variously ignored, disputed or downplayed the golden eagle report and the other deaths and disappearances.

'When so many birds are still inexplicably vanishing, can they really believe persecution is no longer a problem?' I asked. 'Or are they just representing their members?'

'The latter, I'm sure,' said Ian.

'But how can it still be happening?'

'Some people are still living in the Victorian era,' he said simply.

Driving home, I thought of all that we'd discussed, thought again of those questions I'd set out to answer. How successful have kite reintroductions been, and are we really winning?

If you go on population statistics, there can only be one conclusion: the projects have been a huge success. Between 1996 and 2001, 103 chicks were released in Central Scotland. They have long since spread across our borders into neighbouring Tayside. A handful have made it on into Angus. Now there must be more than a hundred pairs breeding between the three regions. Each year their number expands, their range grows larger.

Most of the reintroduction projects have been just as successful as ours; some have even been more so. Between 1995 and 2016 the UK kite population increased by 1,450 per cent, making kites one of the fastest-growing bird species in the country for that period. As of 2017 there were an estimated 6,000 breeding pairs in the UK and Ireland. Now a quarter of the world's red kite population lives in these lands. As the former RSPB red kite officer Kevin Duffy told me, 'Although you'll find some red kites in places like Africa and the Middle East, they're almost exclusively a European bird. When we started the reintroductions, we used to joke that one day we'd end up exporting chicks back to the Continent to help revive any struggling populations. European numbers were so high then that that seemed ridiculous, but now our numbers are rising, theirs are falling – by as much as 50 per cent in some countries – and the joke has become a reality!'

Duncan Orr-Ewing, the former head of the Central Scotland red kite team, added, 'When I first moved to the Doune area, raptor persecution was so rife you barely even saw a buzzard in the sky. Now they're very common. So are kites. The situation is definitely improving.'

In this age, when so much of Britain's wildlife is declining, the kites represent a rare thing: a conservation triumph. They're re establishing themselves in our landscapes, in our lives and in our hearts. In places like Argaty they've become such a familiar sight it's like they were never gone. Thirty years on, these birds are back, re-established, here to stay.

So Mike was right to say that we are winning. But *winning* is the key word. Kites are faring better than anyone could possibly have hoped, but as long as people continue to kill them our success will remain a qualified one. We may stay in a winning position, but we will never be able to say we have won.

18

Winter Roost

One of Britain's rarest birds of prey has bred successfully
in Scotland for only the second time this century, the
Royal Society for the Protection of Birds and the Joint
Nature Conservation Committee has announced ... Seven
wild red kites have been raised this year in Scotland
compared with one fledged last year.

'Kite Makes its Mark', *Lennox Herald*, 30 July 1993

Late January. In the fields below Argaty House, the light was
beginning to fade. I leant against an old dry-stone dyke, my eyes
turned to the sky. Above the ghostly building, thirty kites were
circling.

On a set date each year raptor workers across the world station
themselves beside red kites' winter roosts to count the birds as
they fly into the woods for the night. The collated figures tell us
how large each country's winter population is. Ross, a ranger we'd
taken on to help with the project, had joined me for the count.

There's something not only serene about the birds at this hour,
but strange too. For a while that evening I struggled to think what
it was. Standing in the silence, I watched, counted and pondered.
Then it hit me. The strange thing *was* the silence. Kites are so
seldom quiet. Wherever I am on Argaty I can tell where they are. I
may not be able to see them, but if I wait long enough I always hear

them. Now their conversation was over, and there was something so beautiful in that, so companionable. It was the comfortable sort of silence Sarah and I might share, the sort you can only have with the people you know and love.

Night was coming, the day was slipping away, taking with it all of the birds' colours and markings. Before long, their gently arching wings and great forked tails would be the only defining features they had left.

In that failing light Argaty House looked pitiable. Darkness emphasised the destruction the flames had wrought. With the scorched walls reduced to shadow, the empty spaces where the windows and roof should have been stood out so horribly. It was a tragic sight. I thought of Charlie, wondered what he'd make of it. I'd finished his journals recently. A shock had awaited me when I reached the final pages.

> 1937. The year opened with a visit from my Father and Sister and later Aunt Adeline came from Newlands. The second week brought us the tragedy at Lerrocks when . . . my Farm Manager's wife, in a spasm of temporary mental derangement took the life of her elder daughter Maysie and attempted to take her own. Her immediate removal to Stirling Royal Infirmary and the burial of her child facilitated the handling of the problem which remained, which after a great deal of vexation and heartsearching and some proper handling resulted in the lady's release from the criminal charge on grounds of insanity and her detention during H.M. Pleasure. Further negotiation enabled her to be detained at Murthly Asylum at her husband's expense. His excellent behaviour, courage and fortitude deserve to be recorded.

Lerrocks, where these dismal events occurred, is the house I grew up in.

'Did you know about this?' I asked my mum the following morning.

'Yes,' she replied. 'I was told that she was suffering from depression. Tragically she drowned the girl in the bath.'

A pause. Then, as if anticipating my next question, she said, 'You and your sister were young when we moved into the house. I wanted it to be a happy home, without all that bad history hanging over us.'

I understood that.

Later, rereading that paragraph, reflecting upon Charlie's choice of words – 'vexation', 'heartsearching', 'proper handling', 'negotiation' – I wondered whether a word from him to the right people had spared the woman from the gallows.

Charlie says little more about the sorry incident. He writes of trying to 're-arrange matters at Lerrocks', but that is all. The rest of the year is given over to the usual anecdotes: the coronation of King George VI and Queen Elizabeth; a shearling ram sold for £120 at the Highland Show; Hubert, his eldest child, starting school at Harrow, my grandfather feeling lost at prep school without his big brother; a trip to Germany with the Royal Scottish Forestry Society.

On this latter trip he notes, 'The country appears to be happy and contented chiefly perhaps for the establishment of a settled regime in place of a former chaos not perhaps love of the regime. Much evidence of martial activity and the number of cruising aeroplanes were phenomenal.'

An unpleasant chill came over me as I read this. In two years' time war would break out, in seven Hubert would die in Normandy. How sad to read those lines, knowing what would happen next.

Like most others, 1937 closed with a spate of parties and the customary Christmas cine film reel. I flipped the page, ready to read on, but the remaining sheets were blank. The journals ended there.

'What made him stop?' I asked my parents.

'Those diaries cover the period when he was modernising the estate,' Mum said. 'Maybe once that was done, he didn't have much to write about.'

I couldn't believe that, couldn't see how such a thoughtful, articulate man, so dedicated to the recording of life, could ever have run out of things to say. But there it was. What was in these journals was all I would know of my great-grandfather's life.

'Here come more kites,' said Ross.

I returned my attention to the sky. A group of stragglers were moving in from the west. They appeared over the Drum and hurried in to join the others.

At this hour kites seem to be one bird rather than many. One consciousness guides them, one consciousness drives them. The whole group swirled around for just a few moments longer then reached a silent agreement. They were ready for sleep. One by one they folded their wings, dropped towards the trees and were gone.

I continued to stare up, watching the sky I've known my entire life, a sky suddenly empty of kites. How naked it looked without them. How forlorn. How wrong. The thought struck me: the sky I knew as a child was as empty as the one I was gazing at now. Once there were no kites here. I rolled that idea over in my head again and again, but still couldn't make sense of it. I couldn't imagine a time before the kites.

19

Losses

A bird which has been extinct for more than a century
could be on the way back . . . red kites have been reintroduced into two top secret locations in England and
Scotland. And this year they had their best ever breeding
season with 28 pairs rearing 50 chicks.

Newcastle Journal, 23 July 1994

February came, spring lay on the horizon. Buoyed by the thought,
I began planning what we would do in the months ahead, the
events we'd organise, the projects we'd take on. Some weeks earlier
we'd co-hosted a talk with Scottish Badgers titled 'The Place of
Badgers in the Scottish Landscape', which had been so popular
that we'd had to turn people away. Bob Elliot, former head of
investigations for the RSPB, now the director of the animal cam-
paigns charity OneKind, had just signed up to speak at Argaty in
June. Philip Price was returning in July, this time to deliver a talk
on eagles. Others had indicated their willingness to work with us.
The new year was shaping up to be an even better one than its
predecessor.

And then, entirely without warning, things went awry.

Early in February the pine marten I'd been feeding for months
abandoned the site. One night she was there, the following night
she'd gone. Each morning I returned to find the food untouched.

Why had she left when there was a free meal here for her? Had she been pushed out by another marten? Had she decided to explore the many woodlands around her? I couldn't work it out.

The loss hit me in ways I wouldn't have anticipated. I had been so sure that she would stay, so certain that we'd open a marten hide. The idea had become an obsession. Since childhood I've always been a collector. For me it is never enough to have a passing interest; if something piques my curiosity, I have to immerse myself in it. Having shown no interest previously, I turned fanatical about football in my early teens, learning everything there was to know on the subject. Later music became my craze. Whenever I discovered a new singer or band I had to listen to every note they'd ever recorded, over and over. I drove friends and family mad. One thing always linked my obsessions: a sense that what I had was never enough. Something always compelled me to dig deeper, discover more.

It was hard not to see martens as the latest manifestation of this mania. Ever since I'd first seen one on my stealth camera, way back in 2017, I'd wanted them. I'd read everything I could find on them, watched hours of online footage, marvelled at the skill of those awe-inspiring people who'd set up marten viewing facilities. What strange and mystical powers did they possess? What level of fieldcraft was needed to get so close to these skittish, secretive creatures?

Another irrefutable truth was that I'd measured myself, and the progress I was making, against the martens. Through a mix of ineptitude and misfortune I'd first lost the breeding pair, then the lone marten that had visited the squirrel hide. This time around things looked different. I'd been cautious. I hadn't rushed to set up lights or pop-up hides. And this marten had stayed. That fact was like a badge of honour. I may not have displayed it publicly – earlier failures had taught me the folly of bragging too soon about success with these animals – but I kept it about my person at all times. Simply knowing it was there was enough to boost my morale.

But now the marten was gone, any thought of a new hide vanishing with her, and I felt lost, for if having a marten signified progress, then losing one surely symbolised the reverse.

Part of my identity was bound up with that marten, and one night that part took a walk into the woods with her and didn't come back.

*

As the old saying goes, the devil finds work for idle hands. A lack of projects led me not to mischief but to other forms of darkness. In those low, empty moments after the marten walked, old, unwanted doubts came back to plague me.

There are things in my life that I struggle with. At various points over the past years I had felt out of my depth in the wildlife community. I'd worried about how I'd forge a path of my own after Mike left and doubted my ability to see through plans and projects. But at this particular fork in the road, it was my discomfort with the killing of animals on Argaty that consumed me.

One of my clearest childhood memories is of the estate gardener, a nice man whom I loved dearly, driving his little tractor around the walled garden, towing me and my sister in a trailer behind, then stopping, disappearing for a moment and returning clutching a dead rabbit. Presumably it had been caught in a snare. He threw it into the trailer beside us and drove on.

'He didn't know it would upset you,' my mum told me that night. 'The older generation don't think that way.'

I remember the awful feeling when I first understood what happened to the animals on our farm, why they suddenly disappeared on a lorry and never returned. I never came to terms with that knowledge. Like most people, I simply trained myself not to think of it.

And I remember the shoots. As a teenager I would go out beating for pheasants and grouse, flushing them from cover, pushing them towards the guns. But as I grew older, I became more critical. I heard the quips – 'Great shot!', 'Right between the eyes!' – and I felt completely removed from the people around me. Try as I might, I couldn't see why they were doing this. More often than not I ducked out and sat in the host's house with the old people, small children and retired gun-dogs waiting for the day to end.

I come from a long line of farmers and shooters. Argaty was purchased for its sporting and, to a lesser extent, its farming potential. How to feel about that? One part of me has no problem with non-intensive shoots. This side of my brain is practical. It sees the potential benefits of well-managed shooting to conservation. A habitat managed for pheasant or partridge shooting can also benefit other wildlife, even if the birds themselves may be detrimental. (An estimated sixty million pheasants and partridge are released in the UK each year. Although there has been no official assessment made of this – which in itself is worrying – these non-native birds, when released in such volume, must be outcompeting native species for food. They must surely also attract foxes, martens, buzzards and other predators, which in turn will predate other species.) Setting my concerns aside, an estate with good woodland cover will provide shelter for game and non-game birds alike. Hedgerows and shrubs make good nesting places for both too. Wildflower-rich field margins attract insects, which then attract hungry birds. Would it be better if landowners maintained such habitats and didn't pack them with non-native species? Undoubtedly, but the countryside must run as a business and shooting can bring money in. For all of my reservations, I do acknowledge that.

What of the other part of me, the side that loves animals and abhors the idea of violence against them? That side, as you might imagine, hates shooting.

For years, the latter had been the dominant twin. I may have been lost, without any real clue as to who I was and what I should do with my life, but I clung to one fact: I was the first male in my family for generations who didn't shoot. And I was proud of that, proud not to be part of that set.

But it's hard to define yourself by what you are not, and harder still when you choose a life-path that blurs those lines. In all of those years that I nursed my resentments against shooting, I gave little thought to the morality of meat production. In recent times that has changed. I've begun to question it and my role in the chain, to wonder whether I'm really any different from those who shoot. And during that February of 2019 it was a question that I found impossible to ignore.

Am I any different? Am I *really* any different?

Ask a question enough times and it becomes all that you can think of.

I had jobs to do, things to organise, but something had come over me and I couldn't do them. I felt just as I had in the autumn of 2017 when Mike left, those first squirrels and martens had deserted me, and nothing was working. But it didn't make sense. So much had happened since then. We had found the squirrels, I'd grown used to Mike's absence, we'd had media attention, won an award. We were succeeding to a degree I could never have envisaged. So why was that question still eating away at me?

Self-analysis is something I try to steer clear of. Those who can't do that, who scrutinise themselves too closely and too often, never seem like the happy ones. I try to avoid going down these rabbit holes, try to remain on an even keel, but sometimes I can't help myself.

Why, I wondered, was I so intent on having this marten project, on saving the next thing? I thought back, wondered what had first drawn me to saving squirrels. Was it just about Mike, about convincing him to stay? If so, why choose *this* road? Why did saving things seem like the only means of changing his mind? Could it be that this was about something other than Mike – was this really all about me? Never in my earlier life had I shown any real interest in saving anything. But for two years now I'd been obsessed with it. Why? And why did it now feel like such a crisis that a marten had gone, and one tiny plan had failed?

I thought further back, to that final lambing season, and asked myself if that was when this saviour complex was born.

That day, in the spring of 2017, I'd invited two friends to join me for an afternoon's work in the lambing fields. Sheep, like humans, often struggle to give birth. If a lamb doesn't present in the correct position it can get stuck, causing massive problems. In many cases the lamb won't survive. In some extreme circumstances the ewe can die too. Farmers have to check their flock as often as possible, ready to deliver lambs and minimise losses.

Although our lambing only lasted a few weeks, the days could be long. I often invited friends to join me; they kept my spirits up.

'Let's hope we find a ewe that's ready to pop,' I said, as my friends and I drove towards the fields that afternoon. 'One of you can deliver the lamb.'

I wanted them to see some excitement, to understand something of the intensity of the season and of my life at that time of year. I was, without doubt, showing off.

As we made our first round of the field I saw what I'd been looking for. A ewe stood apart from the rest of the flock. She pawed the ground, rolled onto her side then stood again, groaning. She was trying to give birth and she was struggling. We approached her slowly, got within reach. With a quick flick of the wrist I shot my crook beneath her, hooked a leg and flipped her off her feet. Now, lying prone, she was in a position for us to lamb her.

'Let me see where the lamb is,' I said, rolling up a sleeve. 'Then you can take over.'

Like anyone who's ever worked in a lambing field, I've lost lambs. It can't be helped. You can't be there all the time. Your presence unsettles the sheep, causes complications. That's just the way it is. Sometimes you arrive too early and the ewe goes into labour just after you leave. Sometimes you arrive too late and the lamb dies before you can pull it out. Sometimes, you time it perfectly. It really is luck of the draw.

No matter how badly the lambs have been stuck, I've always managed to pull them out eventually. This one was different. The mother was pushing hard, but the lamb had twisted inside her and one of its legs was caught on her pelvis. I knew right away that my friends would play no part in this delivery.

I tried and tried to move the lamb into a position where the leg would come free. I pushed it, pulled it, turned it. It wouldn't budge. My hand was deep inside the ewe, pushing towards her stomach. With each contraction she crushed my fingers. The pain and frustration grew. At the back of my eyes I felt tears building.

Forced to admit defeat I lifted her into a trailer and took her back to the farm. Dad was waiting for me. We worked for over an hour, trying to manipulate the lamb's position. The ewe lost blood, a lot of it, but at last we managed to get a rope inside her and hook it round the lamb's leg. Then we pulled as hard as we could.

The string dug deep into my already throbbing hand. My skin began to chafe and then to burn. More blood dripped from the ewe. Another heave. The lamb moved, its leg came unstuck. Smiling reassuringly to my friends, I braced myself for the final pull.

The leg snapped like a twig. We managed to drag the lamb out, but there was nothing for it, it had to be euthanised. Dad patted my shoulder, told me I'd done what I could.

'Go and get some dinner,' he said. 'I'll deal with it.'

To that point I'd never euthanised anything. I'd always ducked out, let someone else do it. I could have done the same that day, but I didn't. A voice inside me said that I'd never be a farmer unless I took this step. It told me that I couldn't hide behind my father forever. I had been with this lamb, had been the one who'd tried to deliver it into the world. This was the price for failing in that task. So I lifted it, carried it to the back of the shed, steadied my breathing. Taking firm hold of its legs I swung it hard against the wall. Then I did it again to make sure that the job was done.

My visitors went home after dinner and I completed the final field checks alone. It was a quiet shift, all the ewes were at ease, settling down for the night. Normally I'd return to the farm, grab a dram with Dad and finish up, but that evening I sat in the grass, looked out into the evening sky, watched the sun dip behind the shoulder of the distant Gargunnock Hills and thought about the day.

At home that night Sarah poured me a drink. I told her everything.

After a silence that seemed to last forever she asked that most pertinent question: 'Are you sure you can do this for the rest of your life?'

I was. Or at least I carried on pretending that I was. Then Mike told me he was leaving and the wheels came off the wagon.

I am so grateful to Sarah. Nobody knows me as she does. If not for her I would have weathered the post-Mike storm, found my feet again and continued as before. I'd have gone on living that farming fantasy until it was too late to start again.

In all likelihood your opinion of livestock farming will determine your reaction to this story. If you are opposed to meat production, then this will do little to improve your opinion; if you aren't, then perhaps this will read as an example of the sort of thing farmers have to do if they're to feed people.

My position? I do still believe in the best of British farming. I see the good that livestock can do for biodiversity. I know that people need food and that in our climate grass grows better than any food crop. Given the choice of eating local, sustainably-produced, grass-fed beef or mutton, or products flown in from the other side of the world, I think the former must be the more environmentally friendly. I also worry about what would happen to these lands if we stopped farming them. More than 5.73 million of Scotland's 6.2 million hectares of farmland falls into the Less Favoured Area category. Until such point as politicians begin paying landowners to restore lost habitat, the most profitable thing to be done with this impoverished land will be to plant quick-to-grow, quick-to-fell non-native conifers. These trees are useful only to a handful of generalist species and a disaster for those fragile ones, like ground-nesting birds, which need sensitive, considered land management. A Scotland of Sitka spruce is not the country I want to live in. For these very practical reasons I do believe in livestock farming. I believe in what my parents have done for Argaty too. The wealth of wildlife we have here is all the proof I need that they made the right choices. My trouble is that the instant I euthanised that lamb I lost the stomach for the job.

Now, almost two years later, I found myself in this empty moment with nothing in front of me, no exciting project to throw myself into, and I saw myself clearly. My motivations were laid bare and I understood that I had never made peace with what had happened that day. If I wished to keep some animals alive now, it was because there were others that I didn't save. There was something else too, an uglier truth, buried deeper. The fact was that I didn't feel much for that lamb. Euthanising it had *not* been hard. The moment *didn't* haunt me. I know that it should, that a younger me would have been devastated by it. But I wasn't. I'd been in the game too long, seen too many things die, found rational reasons to justify this life, this way of producing food. I hadn't quit farming because I was haunted. I'd quit because I wasn't, because I saw how far I'd gone from the person I thought that I was. And if I could kill something and feel so little about it, I didn't know if I was any different to the shooters, whose choices I had always criticised. If I was capable of that, I didn't really know who I was.

20

A Turning Point

I understand from RSPB sources that up to forty young
kites have been successfully reared in Scotland this
summer. With this new release, it can be safely assumed
that red kites, until a few years ago, in British terms at
least, the almost exclusively Welsh residents amongst our
raptors, will become increasingly commonplace in many
parts of Scotland.

'Keith Graham's Country View',
Perthshire Advertiser, 23 July 1996

A conversation with Peter Cairns helped to banish these dark
clouds. Pete is a conservation photographer and a director of
SCOTLAND: The Big Picture (STBP), a team of communica-
tions experts who advance the case for rewilding in Scotland. In
2018 I'd seen him give a spellbinding presentation of their vision
for our countryside to the Edinburgh Photographic Society. At
that stage I was new to the idea of rewilding (a landscape-scale
conservation movement aiming to restore and repair ecosystems,
connect wilderness areas, reintroduce and protect top predators
and keystone species). Thanks to Pete's lecture I became hooked.
When he subsequently asked Argaty to join STBP's list of sup-
porters I couldn't have been more pleased.

In a telephone conversation that lasted nearly two hours we

discussed so many topics: the importance of scavengers like kites to a healthy, functioning ecosystem; the need for more wild places where nature can thrive; the problems caused in Britain by our refusal to countenance the reintroduction of missing predators.

'Where does the fear come from?' I asked. 'Why do you think we're so afraid of bringing these animals back?'

'I think Britain, more than most places, is afraid of change,' he replied. 'We haven't had to co-exist with large carnivores and we're terrified of altering that.'

'But we need them back, don't we?' I asked.

'Without doubt.'

'Do you think it'll happen? The politicians seem so against it.'

'There's still a lot of opposition,' he agreed, 'but things are changing slowly.'

'But we don't have time to wait for these changes,' I objected.

Perhaps my tone had betrayed too much impatience, for his instantly changed, becoming more measured as mine became less so.

'You're right, of course,' he said. 'I don't want to patronise you by saying this, but you're a young guy and when progress is this slow it probably seems as though it's not happening at all. But I've been involved in conservation for a long time. Even ten years ago, only a tiny handful of people were into this whole "rewilding" thing. There was a misconception that it was just a bunch of fanatics who wanted to bring back wolves. Now it's on the agenda and people know it's about more than just reintroducing missing animals. It's about restoring and connecting habitat, allowing self-willed natural processes to happen, standing back and letting nature do what it wants. More people know that now, and more believe in it too.'

Our conversation turned again to kites. I put to him what I now considered my pet theory: that these birds are the overlooked sibling in the rewilding story.

'Good point,' he said. 'Maybe they have been ignored. I don't know why. Perhaps we've missed an opportunity there.'

We spoke about British history next and I told him of the stories I'd learned, of Colquhoun, St John and Booth. What, I asked, did the Victorian treatment of predators say about their society?

'I'd hate to put myself in the mind of a Victorian, living in that era of such extreme materialism,' he said. 'Having said that, I grew up in Thatcher's Britain, when teachers literally told you that the only way to get ahead was to shit on anyone that stood in your way, so perhaps our country hasn't changed so much! Maybe we still believe it's our divine right to rule the world and do whatever we want to it. I certainly think that was the Victorian view.'

'With wildlife struggling so badly on our watch,' I said, 'don't you sometimes wonder what future generations will make of us and our treatment of the world?'

'Absolutely,' he replied. 'I think if you started society again and someone said they wanted to use so much of the countryside for meat production, unsustainable forestry and field sports, you'd think they were absolutely nuts. If it seems normal to us, that's only because we're conditioned to it.' After a pause he added, 'And I think our idea of conservation would baffle future generations, too.'

'How so?' I asked, my interest piqued.

'Think of all the time and money that's been spent trying to conserve individual species. We've been slow to understand and appreciate ecological processes, to see the big picture. We need to see nature as a complex web of species interactions rather than a suite of individual components.'

At that moment something happened. How to explain it? I knew that these words were important, that when this conversation was over I'd return to this point and find something there.

We talked on for a bit. As we were about to sign off, I asked,

'Knowing how bad things are with climate change and biodiversity declines, is it too late to turn it around?'

'Possibly,' he said. 'But what are you going to do? Sit and moan or do something positive?'

We rang off and I took the dog for a walk. Leaving the road behind, I travelled cross-country, over seldom-used fields to a never-visited pond. I wanted to be alone, free from interruption, to have time to think.

Human interactions tire me more than they ever used to. The more time I spend on my own, the worse that gets. Nature of the job, I suppose. I still enjoy conversation, still like people (and for avoidance of doubt, I should say that I like Pete Cairns more than most), but I need to regroup and reflect after each conversation.

Finding a nettle-free part of the bank, I smoothed the grass and sat down. A scything wind sliced the rushes, flattened the grass; the pond rippled as though something monstrous was stirring down below. In months to come dragonfly nymphs would emerge from the darkest reaches of that browny-green water, climb the bankside rushes, shed their skins and fully formed dragonflies would take flight; blossoming mushroom clouds of flies and cleggs would fill the air and sand martins would wheel like little guided missiles, picking them off. Geese would nest here, mallards too. But not today. Today the wind was the only thing that broke the surface. Sitting out there alone (save for the dog) in that no-man's-land, I felt like the last human on earth.

I mulled over the conversation, pushed it this way and that as a magpie picks for treasure. I'd called to find out more about rewilding, about the transformative effect that it could have upon our landscape, and to see where kites fitted into that picture. Pete had given me all of that and more.

Think of all the time and money that's been spent trying to conserve individual species. We've been slow to understand and appreciate ecological processes, to see the big picture. We need to see nature as a complex web of species interactions rather than a suite of individual components.

I thought of those words, considered their implications.

What are you going to do? Sit and moan or do something positive?

I considered this too, realised that one rhetorical, throwaway comment had perfectly articulated the challenge I faced. I couldn't continue to ruminate. It was high time I did something positive and for the first time in what felt like forever I knew what that would be.

21

A Fresh Approach

Britain now has 429 pairs of Red Kites – the most for
150 years . . . New figures show 131 pairs in England and
39 in Scotland.

Birdwatching Magazine, January 2001

In my short conservation career to this point I'd engaged in various experiments, attempting to feed squirrels, martens, kestrels and sparrowhawks. The marten feeding was always doomed to fail. I saw that now. As so much of Argaty is woodland, any marten here could roam over hundreds of acres without ever having to leave the trees. Why stay in one place indefinitely, eating the same meals of nuts and eggs each night, when a world of other edible delights lay all around? As for the kestrel and sparrowhawk efforts, they were so unsuccessful I haven't even included them in this book. That left the squirrels as my sole success story. One in four. Not much of a strike rate. My methodology was clearly flawed. I was trying too hard to control the outcome, to make the wildlife do what I wanted. If the specific animal didn't come, the whole experiment failed. It was, as Pete Cairns would attest, a waste of time and money.

Now I was ready for a new kind of experiment. The tactic would be simple: stop trying to manage everything. Create a habitat. See what happens.

As I'd learned more about the worldwide wildlife collapse, I'd grown increasingly concerned about one issue in particular: insects. Recent studies suggest that 40 per cent of global insect species are suffering dramatic declines due to climate change, overuse of pesticides and intensive agriculture. Looking around the gardens in the villages and towns, I couldn't escape the feeling that AstroTurf lawns, monoblock driveways, and a serious lack of plants and flowers had done a lot of harm too. At present, insects are declining at almost twice the rate of vertebrates. What happens to the rest of the food chain if you remove the lowest layer? Visit the supermarket and take a can from the bottom of the pyramid. You'll get the idea.

What struck me as the most troubling part of this story is that so few people seem to have spotted it happening. Small creatures escape our notice, I suppose, yet we rely on many of these insect species for pollination, to break down organic matter, and for weed and pest control. Lose our insects, especially the pollinators, and our whole world could collapse.

It seemed to me that if I really wished both to help with this and to boost wildlife numbers on the estate, something had to be done on the food chain's ground floor. We had to concentrate on the insects. Improve their fortunes and everything else would surely benefit.

At the entrance to the red kite centre is a patch of rough ground where nothing except brambles grows. When visitors arrive, this is the first thing they see and it looks a fright. This would be the site for the new project.

During our dragonfly day the previous summer, Ruary Mackenzie Dodds and Kari de Koenigswarter had spoken with such passion about their love of these vibrant, splendid insects. Their fears for them were palpable, too. Dragonflies (this term is used in reference to both dragonflies and damselflies) were on this earth 150 million years before the dinosaurs and 200 million years before birds. But now, after approximately 350 million years of existence, they were in trouble. The draining of wetlands and the toxic drifts of insecticides and pesticides used too close to watercourses have hit them hard in Britain. Three species have become extinct since 1960, and of the thirty-nine remaining types 36 per cent are in decline. This should worry all of us. Dragonflies tell us so much about the cleanliness of our water and air. If either is polluted, the dragonflies won't stick around. If we are creating a world that is unfit for them, aren't we also creating one that is unfit for us?

Towards the end of his presentation Ruary had changed to a new slide in his PowerPoint and the most amazing construction met my eyes. Some years earlier he, Kari and a team of volunteers had built a pond at the Wicken Fen National Trust nature reserve. I use the word *built* rather than *dug* for a reason. This was a raised-bed pond, made from wooden sleepers. They screwed them together, put a pond liner inside and filled it with water and dragonfly-friendly plants.

Ruary and Kari really sparked something in me. I couldn't believe how little attention I'd paid to dragonflies previously. I watched the common blue, azure and red damselflies, the four-spotted chaser and common darter dragonflies as they fizzed around the water, landed on the bank, then spiralled off again and I was captivated. They really were incredible. Ruary's great hero Norman Moore described dragonflies as the birder's insect, and I understood that. The way they hovered, spun in their tracks, then spontaneously exploded into flight was very raptor-like; their abilities inspired the same awe. Both were capable of acts of physicality of which we humans could only dream.

One thing that truly stunned me was how short the life of a dragonfly is above water. Most species spend years in the underwater larval stage and mere weeks as free-flying adults above the surface. And that was if they were lucky. Sometimes, said Ruary, a sharp-sighted bird eats the dragonfly nymph at the very moment it emerges above water. Bad weather can further shorten those already short lives. Neither dragonflies nor their prey can easily fly in rain, and dragonflies can starve to death in prolonged periods of wet weather. There seemed a real tragedy to the existence of these brilliant creatures. Theirs truly was the briefest of flames. It wasn't hard to see why Ruary and Kari were so devoted to their care.

The things I learned that day remained with me. The image of that raised-bed pond lingered, too. In recent months our volunteer numbers had grown and, with nothing else on my hands, no other work to give them, now seemed the perfect time to build one of our own.

We began construction in late February, quite the worst time for any form of outdoor work. Rain and snow were our constant companions. Thick slabs of clay soil glued themselves to our boots. Walking without falling was nigh on impossible. Then frost came and made the liner brittle. It ripped, and the pond spilled its guts, sending several tonnes of water tumbling to the road below. Mourning the death of the old liner, awaiting delivery of a new one, we were forced to break off from work.

March arrived. In cities around the world pupils skipped school and joined demonstrations demanding political action to tackle climate change. The following month a group of activists from the group Extinction Rebellion occupied prominent sites in London, adding their voices to the protests. The Scottish Government, UK Government and Welsh Assembly responded, declaring climate emergencies and promising to cut net carbon emissions to zero (by 2045 in Scotland's case, 2050 for the rest of the UK). At last, at long last, the environmental crisis was front-page news.

Towards the end of the month I realised an ambition. I saw a pine marten. Although I could very easily have visited a marten hide had I so wished, some stubborn part of me refused to. Either

I would see one on Argaty or not at all. When the latest one had abandoned me, I'd resigned myself to failure. But then we began setting grey squirrel traps again, and one night, in the woods above my house, I caught her.

Since the turn of the year we had been working with the SSPCA, releasing hedgehogs that they'd rehabilitated. The hogs, I soon learned, had a taste for the maize we used in the squirrel traps. Time after time I found them curled up in the cages, waiting to be released. Trying to get a balled-up hedgehog through so narrow a doorway is about as easy as removing a ship from a bottle.

When I approached the trap and saw a closed door that morning, I fully expected to find another hog inside. What a shock to discover a marten instead. For several moments I sat taking her in. Her eyes, slanted like a Siamese cat's, were exotic and so very beautiful. The feral, fruity smell of scat emanated from her. Here, at last, was the enormous, bloodthirsty predator of historical legend, the creature that had stalked my dreams. The reality was nothing like the myth. She wasn't that big. She didn't look *that* fierce. In truth she looked as any other animal would under such circumstances: very afraid. I felt ... what exactly? Simultaneously wowed and disappointed sums it up, I think, though the latter feeling had nothing to do with the marten. She was wonderful. Rather I was disappointed in myself. Even though the nonsense written about kites had taught me to have little faith in the natural historians of old, I had still given credence to their accounts of martens. In my mind I'd been tracking something truly fearsome these past months instead of this small and very pretty animal.

She sat trembling, watching me all the time. Taking up a handy stick, I began to pry open the door of the cage. As if from nowhere a slashing paw shot out. The stick flew clean out of my hands. In the long minutes that followed, while I tried to release her and she repeatedly lacerated my work tools, I made a mental vow. Never again would I complain that hedgehogs were hard to set free.

Eventually I succeeded. She shot from the cage like an acrobat fired from a cannon. What a strange sensation to watch this scared little animal disappear through the bushes and out of sight. For nearly two years I'd been obsessively chasing martens, desperate to see one, to gain its trust, to set up my damned hide.

When I think about it, each success and failure seemed so much greater than it was because I believed it all symbolised the journey I was on, my quest to prove that I was capable of living this life. But if the search for martens had taken on a metaphorical dimension, then perhaps letting this one go also had a deeper meaning. Since that day I haven't felt the need to collect anything. I've rid myself of this compulsion, shouldered since childhood. After all these years I've finally let it go.

The weather improved, a new liner arrived by post, we started over on the pond. So as to allow amphibians access to the water, I jumped on our old digger, a groaning, rusting, chipped-paint dinosaur of a machine, and moved heap upon heap of earth, creating a ramp that rose to meet the southern wall. Whenever a tractor was available I borrowed it, lifted a water pod on the front spikes and added its contents into the frame.

Kites flew over each afternoon, making their way to and from the feeding area. Once in a while a sparrowhawk appeared too, swerving maniacally round trees and bushes like a racing driver taking each corner on two wheels. The hawk had been seen a lot at the hide recently, suddenly breaking cover to ambush sparrows and tits. Its presence did nothing to deter the kites, however. The goshawk, which can kill almost anything, is the only smaller hawk that might frighten them.

As Ian Carter details in *The Red Kite*, there is some interesting history concerning kites and non-native hawks. These birds have a bloody past. In medieval times, particular honour befell the falconer who took down a kite in front of his king. No British bird proved up to the job, however – the kite was too accomplished a flier – so our men had to import lanner falcons from Africa. They lured kites in by first catching owls, tying fox brushes to their tails and setting them loose as bait. When the kite came near, the falcon struck. Stories abound of rival French and English falconers competing to kill a kite in front of the royals of their countries. James I in particular was so desperate to get one over on the French that he spent almost £1,000 on a cast of birds. Unfortunately for him, when the time of the demonstration to his enemy came, disaster struck. Seeing the falcon coming, the clever kite shot skywards, off into the clouds. The falcon gave chase and neither was ever seen again. The legacy of those hunts is still apparent in Europe today. In France kites are known as *Milan royal*, in Spain they're *Milano real* and in Italy they're *Nibbio reale*, the adjective 'royal' alluding to this bloody past. We'd had many visitors from these countries already in 2019, but thus far none of them had known the stories behind the names. Perhaps that was just as well.

After many long days we got the pond filled. What tedious work it had been. Had we a hosepipe long enough we'd have filled it from the mains and finished in a day. Instead we'd been forced to use water pods, which we lifted with the loader tractor. This was definitely the poor man's choice, the slow way of doing things, but now the job was done and all we could do was wait and hope that something found it. Of all the 'somethings' that might come, my biggest hope was, of course, for dragonflies.

A lull followed. I began to consider other ideas that fitted our new habitat-centred ethos. A growing number of tree sparrows had been spotted in the hawthorns around the kite hide. Using old offcuts of wood scavenged over the years by Mike, left gathering dust since his departure, we began to build nest boxes. The end result was very pleasing. On every fence post a box was attached. We'd created a little housing complex, a tree sparrow village.

Our next project took us below the farm, to a four-acre field which was long since rendered useless for agriculture. Years ago we used to keep pigs in there. They'd been gifted to Dad for his birthday, but we were in no way equipped to house them. They bred like rabbits and always found some means of escape. We seemed to spend half our lives herding them back into the field, only for them to break out again. Often they sat on their young and crushed them. Occasionally they even ate them. I couldn't stand that. In the end I begged my parents to bring our brief trip into the pig farming world to an end and they agreed.

The pigs had rotavated the soil, turning the seed bed over, bringing up a jungle of dockens and little else. Neither sheep nor cow had grazed it since. Another unused spot, another site for an experiment.

I had just finished reading John Lewis-Stempel's *The Running Hare*. Stempel, a farmer, writer and nature-lover, was dismayed by the intensification of arable farming and the consequential decline in the animal and plant life he remembered from his childhood. Eschewing the combine harvester and the weed sprayer, he used traditional methods to sow a wheat field, replete with wildflowers. From the insects in the ground to the kestrels in the sky (and, of course, the running hares), so much wildlife came to this one field.

I knew we couldn't follow his path to the letter. For a start, we'd have to use machinery, if only to turn the soil over and get on top of the weeds. Our chances of growing wheat in our four-acre docken patch were also slim (and besides, we had little use for a small wheat crop). But wildflowers? Surely we could grow them.

In April, Dad took the plough to the field, churned the whole mess up and sowed it with wildflower seed.

The pond was built, the bird boxes were up, the seeds were sown. Habitats were there, waiting for occupants. I crossed my fingers and hoped they'd come.

22

Ospreys

A bird of prey which was hunted to the verge of extinction has been seen in Edinburgh for the first time in 300 years.

Reports of a red kite flying over the capital and circling Holyrood Park have left experts believing the 'magnificent' creature could be making a comeback to the city . . .

Julian Warman, the Scottish Wildlife Trust's conservation manager, said: 'This is really excellent news. Red kites are one of Scotland's conservation success stories and the fact that these birds have been sighted in Edinburgh just demonstrates that the re-introduction programme has really worked.'

The Scotsman, 21 June 2005

Before I knew it the nesting season was upon me again. Could it really be a year since Keith and I first began our springtime rambles? So much had happened in that period, and yet the time had passed so quickly.

Back to the woods I went, ready for another nest-hunt. Bluebells were out. Daffodils too. Orange-tip butterflies flickered between them and bumblebees idled past, tubby bellies grazing the clover-covered ground. I followed the deer tracks, the long, wide badger runs, the dried-out beds where streams used to flow. Temperatures were once again soaring. The water was gone.

Often our new ranger, Ross, accompanied me on these trips. I was supposed to train him this year so that he might apply to join the Raptor Group in the autumn, but he needed little tuition. Some people have an eye for the job. We found the buzzards' nests, found the kites', noted their locations on a map.

The more time I spent here, the more my efforts were rewarded. With each visit my ability to look and to listen, to read the land and to hear it, grew; the hunt became easier. For the first time the abilities of people like Keith, for so long magical and mysterious to me, began to seem a little more human. Now I saw them for what they were: a return for years of investment in the natural world.

This season the kites were up to their old tricks again, thieving assorted items, who knew where from. Through binoculars I spotted black silage wrap in one nest; a punctured purple balloon lay on the ground beneath another. I hoped it had burst before the kite gathered it. The bang, I fear, would have provided quite a shock!

One day, out on my own, the most amazing thing happened. En route to check a buzzard pair I passed below one of the artificial nests Mike and Keith had built so many years earlier. Two birds leapt into the air, screaming in protest. They seemed a mix of kite and buzzard, with the former's wing shape, the latter's tail. They were much bigger than a buzzard, however, and their wings, though only a fraction longer than a kite's, bent far more sharply at the carpal joint. Their call – a 'hoo-hee-hee-hee-hee' sound like a dog's squeaky toy – was unmistakeable. They were, without doubt, ospreys.

The following day they went to work, plundering the forest. Alighting in trees they groped at branches and twisted until they snapped. Their great snaking talons, made to grip the largest of fish, seemed designed for such woodland sabotage. They carried their takings back to the nest, added them to the stick pile; the walls of their house grew taller and taller.

Later that week I took Sandra, one of our new volunteers, to watch them. We adjusted our binoculars, found the nest. Nothing was visible.

'Damn,' I said after several minutes. 'Sorry.'

'Wait,' she said.

From the southern skies a shape had emerged. A bird, flying fast in short patting wingbeats.

'That's him,' Sandra said. 'And look, I think he's carrying something.'

She was right. In those hard, branch-breaking talons a fish was writhing.

'And look there,' she said, binoculars turned onto the nest.

Over the mound of twigs a head – white, with a badger's stripe and fuzzy duck hair – had emerged.

'There she is, too,' Sandra said. 'Must've been lying low this whole time.'

The female osprey stepped forward and perched on the edge of the nest, proud as a ship's figurehead. Her partner soared towards her, landed, and together they tore into lunch.

'Yes!' I shouted, jumping around as if celebrating a goal.

I remember Mike giving an account of the first time that a juvenile sea eagle visited Argaty. He'd watched it flee from the feeding area, a squadron of angry kites circling like wasps at its tail. 'Two species in the sky at the same time,' he'd said. 'Both formerly extinct in Scotland, both right in front of me. It was pretty cool.'

A few days later I enjoyed a similar view. Rowan and I were playing in the garden when a shadow swept over the ground. The next second, another passed. We looked up. A kite and an osprey were soaring over our heads. I had to agree with Mike's verdict. It was indeed pretty cool.

'Brilliant!' said Keith, when I told him the news. 'After all this time, waiting, hoping. Maybe this'll be our year.'

Maybe it would. Let the wildflowers grow, the birds find our nest boxes, the dragonflies find our pond, the ospreys have chicks. I crossed more fingers, said more prayers.

23

Gill

The Scottish Red Kite population has reached its highest
level for 200 years, following one of the best breeding
seasons in the history of the species and an ongoing
reintroduction project centred on the skies around
Aberdeen.

According to a new report by RSPB Scotland, the
Scottish population of the raptor now stands at 122
breeding pairs, while the UK population as a whole has
grown to an estimated 1,200 pairs.

The Scotsman, 15 September 2008

More excitement was to follow in April. I received a phone call
came from Colin Seddon, manager of the SSPCA's National
Wildlife Rescue Centre.

'Do you have any ponds on Argaty?' he asked.

'Twelve,' I replied, mentally discounting the raised pond.

'How would you feel about releasing a rehabbed otter in one of
them?'

You already know my answer.

The otter, named Gill (as in a fish's gills), had been handed in
as a cub the previous year. Floods had swept through the Scottish
Borders and she'd been carried away, forever separated from her
parents. The SSPCA had been caring for her ever since, rearing

her, preparing her for release. In that time she'd become a minor celebrity, featuring in *Born To Be Wild*, a BBC documentary which followed the work of Colin and his team.

He brought her out to us one afternoon. Otters can be ferocious killers. I knew that, but to look at Gill I couldn't quite believe it. At less than a year old she was not quite fully grown and hunched in a corner of the carry-cage she looked smaller still. A broad white mark like a moustache stretched across her muzzle. Her eyes were dark and alert, her whiskers were as long as a cat's. I thought her very beautiful.

Below an upturned dog bed we placed three large rainbow trout. Their brilliant scales gleamed like petrol in a puddle. The bed was there to keep the kites, buzzards and herons from the fish. Beside it, Colin set a wooden dog kennel.

'It smells of her,' he said. 'She thinks of it as home.'

Like the proverbial rat up a drainpipe, she bolted into her wooden hutch the instant he opened her cage door. After several minutes, during which we waited, silently, it became clear that she would not come out.

'She'll probably hide away till it gets dark, then swim off,' Colin said. 'The fish are there if she needs them, but females are so independent my guess is we'll never see her again.'

We tied a stealth camera to a post, trained it on the kennel and left her in peace.

In the early hours of the evening, she took her first steps as a free otter. Like a criminal looking out for the police, she spun her head, left to right then back again, as she slunk across the bank. The lure of the water must have been great, for she didn't so much as look at the fish. She just ran, dived and disappeared. The camera caught all of this.

The next morning the fish were still there. Gill had gone; she'd chosen the wild, and it appeared we would play no further part in her story.

A strange surprise awaited me the day after, though. Lying on the dog bed, fat lips gaping, wide eyes staring skyward, was the severed head of a trout. It looked like the sort of omen mafia gangs send one another in Hollywood films. Beneath the bed the grass was bare, the other fish had gone.

The camera, I noticed, was slumped at an odd angle. Had I really pointed it straight towards the ground? Surely not. The first few videos left me none the wiser. A heron sashayed past. A buzzard landed on the bed, sniffed the air; it knew there were fish but couldn't see where. On it flew. The camera remained on target. Then, in the early hours of night, a sleek, slender body emerged from the water. I peered closer, bringing my face to within millimetres of the camera screen, and to my delight I saw a pair of sparkling dark eyes. Beneath them was a thick white moustache. Gill disappeared off-screen for several seconds. And then a large, slick-haired ottery backside appeared in full-frame, gyrated against the camera and sent it tumbling.

She returned every night after that, always at the same time.

'It's a first for us,' Colin Seddon said, as we sat in my parents' kitchen, looking through the videos. 'We've had some otters that couldn't adapt to life back in the wild. You may see them again if they're found and returned to the rescue centre, but we've never had follow-up footage of a successfully released otter.'

'She *does* look healthy, doesn't she?' I asked nervously. 'You don't think she's dependent on the food?'

'Her coat looks good, she's running around, playing on the bank. These are all important signs. Remember, too, that otters need to eat huge amounts to survive. What we're giving her is nowhere near enough to keep her going on its own. She must be getting frogs, grubbing for worms. If she was struggling to hunt, we would see it in the videos. This is great to see!'

'That was really touching, wasn't it?' Mum remarked after

Colin had gone. 'When you think of how many animals he must have released back into the wild, and he was still so excited to see those videos. It tells you a lot. This is so much more than a job for him.'

Gill's feeding habits were strange. Some nights she gobbled the whole fish, others she took just a nibble. I got used to finding heads proudly dumped on the dog bed. She was like a cat, leaving her prey in prominent places for my approval.

Some weeks later, Keith came to check the chicks in a nearby kite nest, gauging their size ahead of the ringing season.

'Three chicks, Tom!' he called down from the nest. A confused frown creased his face. 'Two big fish heads, too.'

'Circle of life, Keith!' I shouted back, laughing in delight.

'What you on about?' he said, looking at me as though I was insane.

That evening, my filmmaker friend, Andrew O'Donnell, visited and together we went out to watch Gill. I haven't written nearly as much as I should about Andrew. Since he'd first come to Argaty to discuss a feature on our squirrel rope bridges we'd spent hours watching and filming wildlife together. He'd produced a couple of beautiful promo videos for our hides and filmed our squirrels for numerous projects. I confess to being somewhat in awe of his work. What impresses me most is the range and depth of emotion it displays. Some of his films are funny, some sad, some angry. He can change pace effortlessly between these moods. Driving him, I think, is that same panic that burns inside most conservationists when they see how poorly people treat the natural world. Recently the BBC had commissioned him to create his own series. He and his friend Mark would go adventuring around Scotland, filming the best of its wildlife and landscapes. Due for broadcast just before Christmas, the show would be their big break.

That evening we huddled in a pop-up hide on the pond's

southern bank, downwind from the small promontory where Gill normally emerged. The air was mild, seasoned with water mint; the sky was a slowly darkening blue. Over the distant hills a huge cloud washed in like a breaker wave. I followed its path, surprised at how quickly it moved across the horizon and disappeared from view. Had clouds always travelled so fast? I couldn't say for sure. How long had it been since I simply sat and watched the sky? So much of adult life is spent looking ahead, thinking of the next thing that has to be done. We never make the time to do the simple things. I had an idea of what the sky was like, some old memory saved from childhood, but it did not match her physical reality. She was a stranger to me now.

A duck swam past, five ducklings paddling in her wake. While she dived they sped around, drawing Roman numerals and figures of eight in the water. Training his camera on them, Andrew followed their path through the reeds.

'No way they'd be hanging about like this if the otter was anywhere near,' he said. 'She must be moving in from somewhere else at night.'

The heron, a frequent visitor according to my camera trap, floated into view, dinosaur wings billowing like a parachute as it came in to land. Somewhere nearby a cuckoo was calling. I searched the middle distance for the osprey. I'd seen him fishing the river a few days earlier, but he was elsewhere tonight.

'Probably about half an hour of light left,' Andrew said, craning his neck and appraising the sky.

Even on this most clement of evenings, we felt the temperature dropping, the night moving in. The promontory where the dog bed lay was growing harder to see. The heron took off again; the ducks vanished.

Earlier that day we'd set up camera traps along the pond's intake burn. Otter spraint – tar-black clumps of bone-filled dung – lay on many a prominent rock. We'd pointed the cameras at these spots, convinced that Gill would follow the stream in again.

Peering through binoculars towards the burn, Andrew frowned. 'Anything?' I asked.

Bins still to his eyes, frown deepening, he shook his head.

I sighed, began to say something about our bad luck. And then, on the hill beyond the pond, she appeared.

'What the hell are you doing up there?' I whispered, grinning.

Ignoring the burn entirely, she ambled slowly down the slope. At the water's edge she sloshed out of sight. Reverberating ripples spread across the pond. Another splash and she emerged, making straight for the dog bed.

We watched her for an hour, saw her dig beneath the bed with her nose, drag a fish out and begin to munch. What a noise! Her table manners really were appalling. Two of the three fish were quickly devoured. She took the final one out into the water, a feast saved for the wee small hours.

Later, as we were tiptoeing away, the sound of breaking twigs met our ears. Beneath some distant birch trees a dark shape was wriggling. Gill was little more than a shadow now, but from the way she moved we could tell what was happening.

'Having a good play,' Andrew said.

'Isn't she just?' I agreed, smiling contentedly as I watched her roll and tumble on the black grass.

We hefted our kit and continued on towards the car. Gill's thoaty calls travelled with us, joyful sounds scattered like seeds by the cool night air.

24

Back to the Woods

The RSPB is celebrating the anniversary of the reintroduction of red kites in Aberdeen, which hadn't been seen in the city's skies for almost a century and a half.

Following the successful reintroduction, numbers have climbed year on year, with at least 35 breeding pairs established in the North-east last year. During that time Aberdeen's red kites have raised more than 300 chicks in the wild.

Evening Express, 17 July 2017

In May 2019 another satellite-tagged hen harrier vanished without trace in the Cairngorms National Park, taking the total number of known raptor killings/disappearances within the park area to sixty-eight since 2002. In June, two of three red kites found near Kirkcudbright in the south of Scotland were confirmed to have died of illegal poisoning. On it went . . .

Meanwhile, at home we had good news to distract us from the bad. Our raised-bed pond had residents, and what a host of little monsters they were. Backswimmers – spidery little demons with a bite as vicious as a wasp's sting – hovered deathly still, bellies in the air, lying in ambush, then exploding into movement. When they kill they inject venom, liquefying their targets' insides. Then they drink them through their proboscis, slurping them up as a child would drink a milkshake through a straw. Whirligig beetles fizzed like little balls of mercury on the surface, their compound eyes divided so that one half could search above water for prey while the other scoured beneath. Nearby, pond skaters danced like splay-legged skiers, the hairs on their legs serving both to keep them afloat and to detect the vibrations of land-based insects falling into the water. Suddenly they'd slide in for the kill, grabbing and stabbing their unfortunate victim. Most thrilling to see was the great diving beetle, the lion of this jungle. These pebble-sized brawlers appeared on the surface from time to time, stashing bubbles of air beneath their wings then disappearing down, down, down, to wreak bloody havoc in the killing grounds below. Beneath the calm and glassy veneer, a pond is a world of horrors and atrocities, a fascinating microcosm of life and death.

Above the water swarms of flies circled. From time to time a sparrow or tit would swoop in and reduce the pond population by one. I thought of the dragonflies. Might they also come hunting here soon? One thing was certain: plenty of food awaited them.

The swallows were back, moving in and out of the old barns at breakneck speed. When swallows fly south, one gets the sense that the best of the year has gone, leaving only mirthless skies and long nights, but what joy to see them return, to hear their childlike voices chattering in the air. They bring life to the year.

Throughout this late spring heatwave, an audience of hungry flies gathered each day, lurking like starving orphans as I chopped lunch for the kites. If ever I left the meat uncovered, the gloves or knife unwashed, I'd return to find the flies laying their rice-like eggs. As working conditions went, these were far from sanitary, but the swallows took full advantage. Giggling hysterically, they shot over my head, grabbing fly after fly after fly.

Our volunteers had worked so hard to revive Mike's bumblebee garden the previous year and the results were incredible. Giant knapweed – a cornflower that seems a mix of dandelion and thistle – towered into the sky, salvias and flowering currants sprayed sweet perfume into the air, lavender bushes stretched their thick, woody arms wide. The bumblebees – white-tailed, buff-tailed and carder – were out in force. On her days off nursery, Rowan and I walked through the garden, stroking the bees' coarse-haired backs as they got drunk as skunks on nectar. We dipped the pond too, dredging up palmate newts that wriggled and writhed in our hands. Recently she'd turned five. School loomed at the end of summer. How I'd miss these lazy days spent exploring the world together.

Tree sparrows were nesting in all but one of our nest boxes. Over and over their fluffy rumps disappeared into the entrance holes, over and over their brown-capped, bruise-cheeked heads reappeared.

Squirrel kittens left the dreys and took their first clumsy trips to the feeders. Stumping around looking for nuts, they made such a racket that had you closed your eyes you could have pointed to the exact spots where they stood. That same month the adults began their spring moult, shedding their old coats and donning new ones. The hair loss began on their heads and spread down their bodies. How tatty they'd look when the moult reached their tails.

Raptor nesting season was in full swing by this point. We ringed the ravens, ringed the tawnies. One evening a friend, David Anderson, flew a drone over the osprey nest, checking for chicks. Dave is a nationally renowned conservationist who has done so much to help re-establish kites, climbing trees and ringing chicks with a zeal that often borders upon the fanatical. Although he spends much of life travelling the country, monitoring eagles, ospreys, goshawk and more, he is always incredibly generous with his time. That evening, while collecting Rowan from nursery, I'd run into him by chance as he was picking up his son.

'I know how busy you are,' I'd said. 'But would you be free some time for a bit of drone work?'

'Aye', he answered. 'Right now. I've got it with me.'

Two tired, unimpressed children were bundled into their respective cars and off we went.

The drone buzzed like an angry fly as he sent it off through the sky. With fine thumb movements he worked the control sticks, guiding it forwards until it sat hovering, kestrel-like, over the nest. The little helicopter's camera beamed directly to his handheld monitor.

'Can you see them?' he said, pointing to some blurry white pixels on the screen. 'Two chicks and an unhatched egg.'

'Definitely,' I said, though in truth those birds were as clear to me as the hidden image in a magic eye painting. It didn't matter. We had an active osprey nest!

The kite chicks, as was evinced by the haste with which their parents began appearing for food, were growing rapidly. On woodland wanderings with Ross, I found ever growing rings of poo on the ground beneath the nests. Clearly the chicks above us were flourishing. The bigger they are, the further they poo.

We had five nests and ten chicks this year. It might have been six nests, but one had blown out in a gale. In spite of this loss I felt encouraged. Though far from huge, our total for 2019 was at least better than the previous year's.

Mid-June arrived and a small group of RSPB patrons came to see one nest's chicks being leg-ringed. The event was led by Duncan Orr-Ewing. Considering that his contribution to ornithology is greater than almost anyone alive in Britain today, Duncan is unfailingly modest. When introducing himself he opted for a simple: 'Hi, I'm Duncan Orr-Ewing. I'm a colleague of Tom's in the Raptor Study Group and I do a lot of work monitoring red kites.' Only when I prodded him did he admit, bashfully, that it was he who had brought kites back to Central Scotland.

I'd long suspected that without him we would never have had kites in this area. A few months earlier, Kevin Duffy, who'd served for several years as Duncan's red kite officer, had confirmed the theory.

'There were plenty of reasons why reintroduction sites were chosen. All the things you read in the RSPB documentation ... suitable habitat, evidence that the kites existed here in previous centuries, further evidence that raptor persecution was no longer a problem ... All that criteria was important, but the other thing, when you look at it, is that kites tended to be released in areas where they had a respected naturalist to champion them, someone who others would listen to. In the Black Isle you had Roy Dennis, in Dumfries and Galloway you'd later have Chris Rollie, and in Central Scotland ...'

'Was Duncan,' we'd said in unison.

'If he'd been based somewhere else, I'm sure that the reintroduction would have happened there instead of here,' Kevin had concluded.

I was on the verge of telling the group this, but as if anticipating some such compliment Duncan had begun singing the praises of the German ornithologists with whom he'd worked to source the chicks for translocation.

Reasoning that perhaps the more one achieves in life, the less one feels the need to brag about it, I smiled inwardly and decided to say nothing. Later that year Duncan would head to Spain. As part of the LIFE EUROKITE project he'd work to restore dwindling kite populations in Extremadura and Andalucia, tagging and monitoring birds to understand their movements and reasons for mortality, and translocating English kites there to bolster Spanish numbers. If kite populations are revived in Europe, as they have been here, it will be thanks to people like Duncan.

We took our visitors to the nest closest to Gill's pond. Thankfully, the fish heads had gone, although a cloth glasses case and a shredded latex glove were there in their stead.

The nest contained two chicks. Their crops (the muscle pouch on their throat where they store food before digestion) were empty. Smoothing their feathers, Duncan drew our attention to tiny lines that broke the pattern of burnt orange and brown.

'These are fault bars,' he explained. 'Malformations that form in the feather if they miss a day's feeding.'

Both of the birds had these marks. In recent weeks our rainstorms had reached Noah's ark proportions. These conditions always have repercussions. For kites, prolonged heavy rain can be a disaster. The matrix at the centre of their feathers soaks up water and the birds can grow so waterlogged they struggle to fly. On the rare occasions when they do take to the sky, they look miserable, just as you or I might if we went out in such conditions wearing a woolly jumper. Faced with the choice of taking a drenching foraging for food or sitting out the wettest days, these birds' parents had clearly chosen the drier option.

They were circling high above us now, keeping an eye on proceedings. Their chicks would get a decent feed today. The sun was out, the wind was up; there'd be no issues flying for food.

As well as the metal BTO (British Trust for Ornithology) ring, we fitted the kite chicks with large, red plastic rings. This was the first year that these had been made available for use on kites, although similar designs had been fitted to other raptor species for some time. Each plastic ring had a number upon it which would be visible from a distance. When they fledged we'd know which youngsters were visiting the feeding area. Should they later leave these parts, there was every chance that someone would see the ring and report back. I wondered where they'd go, whether they'd return. For the first time since we stopped wing-tagging the chicks, these questions would be answered.

The woods really were alive with the sounds and sights of early summer. From a stand of nearby firs, a blackcap was trilling delightfully, keeping his partner abreast of every movement he made. Foxgloves stretched like octopus tentacles all around. Bumblebees floated in and disappeared within their bell-shaped flowers. A red squirrel came running across a fallen tree, saw us standing there, promptly turned and fled. Meanwhile flies and midges were buzzing around the young kites. Clouds of them were hovering over the nest too, no doubt attracted by the rotting meat up there. It reminded me of a story Keith once told me of a time, during ringing season, when he'd climbed to an osprey nest. Midway up the tree he noticed something white moving slowly down the trunk, a line of ants marching like pall bearers in a funeral procession. Slung across their backs was something large, at least compared to the ants: chunks of fish they'd swiped from the nesting ospreys above. That story blows my mind. Who could imagine such a web of interactions? Who could know that a fish, swimming some miles away, would one day provide this much-needed protein to ants in the middle of a forest? The natural world is complex, interlinked and wonderful in ways we barely begin to understand.

Some feral scent was filling my nostrils. Lifting one of the chicks I pressed my nose to its feathers and immediately regretted it. For all their beauty, these birds really didn't smell pleasant. Given the state of their parasite infested nests it was hardly a surprise. I swiftly set it back down.

Later that night, having scratched a spot just below my waistline for over an hour without ever soothing the itch, I took a cursory glance down my trousers. One of the kites' housemates – a big, fat tadpole-sized tick – had transferred itself to me. The skin would remain inflamed for weeks, my own uncomfortable reminder of the endless connections of the natural world.

25

Answered and Unanswered Prayers

It becomes increasingly hard to stay positive about the situation in the natural world, and devastating to accept that the British Isles are recognised as one of the most biodiversity depleted countries in Europe.

Yet as I sit and watch the birds here, I marvel that this is one of the finest examples of a great conservation success story.

This once despised raptor has indeed returned from its brush with extinction and the Bowser family should be proud of the part they have played in its safe return to our skies.

Polly Pullar, 'Red Kite Delight',
Scots Magazine, August 2018

The hawthorn bushes by the hide were positively vibrating. The tree sparrow chicks had left their boxes. Now they sat in lines on the branches, chittering away, begging their parents to feed them. Every few minutes, as if exhausted from the exertion, their chubby little heads slumped and they fell sound asleep.

High above them a kite glided past, hanging a leg. I thought of the bird Mike had told me about, whose leg injury was always at its most pronounced during nesting season when he constantly had to grab food for his chicks. By the end of the season he was unable to tuck it in when he flew. Was this him?

The kites were behaving strangely. Usually so tolerant of each other, all summer long they'd been fighting like cats.

'Watch this,' I told the visitors as the birds gathered, ready to dive.

While all the others bombarded the food, one kite was moving in from the wings at a very different trajectory. Another kite rose, clutching meat. At the last moment it saw the incoming threat and flipped upside down, bearing its talons. The two collided, locked claws, began to grapple. For a split second they made a perfect mirror image of one another. Then they tumbled and the food fell forgotten to the ground.

Over the course of the day the birds kept sparring. Several of them were at it. This unexpected aggression made for spectacular viewing, although I was at a loss to explain it at the time. Subsequent confabs with Dave Anderson, Duncan Orr-Ewing and Roy Dennis shed some light on the situation.

'What you're seeing is something new,' Dave said. 'With the population growing, there must be more competition for food.'

'With more kites,' Duncan added, 'we're bound to see a range of personalities coming to the fore. If an aggressive bird has chicks, they'll probably inherit their parent's temperament too.'

'The young sea eagles we've reintroduced to the Isle of Wight are getting along really well,' wrote Roy in reply to my email. 'It is so exciting seeing them catching mullet, cuttlefish, bass and bream around the island, while on their travels they have located the best places for rabbits. Like the red kite reintroduction the important factor was releasing them in areas of high and varied food availability. I can see the day, however, when their number grows and we have older, more established birds there. A pecking order, like hens in a run, will probably develop, as seems to have happened with your kites.'

All of this made absolute sense, though it led me to another question: Were the scenes I was witnessing something new or, in fact, something very old indeed? While our numbers are still nothing compared to former times – the Victorian gamekeeper's 'vermin' cull records prove that – there are now more kites in Scotland than there have been for over a century, and at Argaty our small daily food dump must rival the amount of carrion found on farms and in the countryside back then. On our farm we are drawing slowly closer to the natural balance found in that distant past and perhaps we are witnessing different sides to these birds as a result. Could it be that the kite we think we know – mild-mannered, gregarious, cordial with its neighbours – is not the bird that inhabited this country centuries ago? When there were more kites, was there more competition between them? Were they forced to hunt more frequently, simply to survive? Did they fight one another for food? We'll likely never know the answers, and while I'll never subscribe to the distorted and overblown views of kites that our ancestors handed down to us, part of me now wonders if the bird they knew was indeed different to the one I know. We have been watching kites for just twenty-three years, a blink in time. What if we were to fast-forward another twenty-three? I'll be in my late fifties, growing older, not in the winter of my life, I hope, but certainly in the autumn. I'll have changed, for sure. Will the birds?

As I sat pondering this another kite swung past, turning on the breeze at just the right moment, allowing me a perfect view of her wing tags. On her left wing was the customary red given to every kite chick that hatched in this area; on her right a white tag. A second fly-past and I caught sight of the code printed on each. S4, the chick from our CCTV camera nest who fledged too early and had to be rescued, only to be rescued again later when she injured her wing.

'Guess who I saw diving for food the other day,' I said to Keith when next I saw him. 'S4!'

A satisfied grin spread across his face. 'S4!' he said. 'Cocky wee bird! Thought she could fly the moment she hatched. She was like that when we ringed her, too. Soon as she saw my face peering into the nest she tried to fly. I had to put my arms round the nest and usher her away from the edge. I gave her good lectures then, *and* when we rescued her the next week. I told her: "Bloody stay put till you're big enough to go!" Christ, if you'd seen the branch she was hanging from when I got to her you'd never think she'd have survived! It was tiny, and she was barely on it. If she'd been a person she'd have been hanging by her fingernails!'

I repeated what Stewart Miller, the falconer who'd rehabilitated the kite shot in 2009, had told me, that every bird is different, that they all have their own wee quirks.

'Maybe some think they're more grown up than they really are,' I suggested.

'Definitely,' Keith agreed.

A beatific smile remained on his face for the rest of the day. What a feeling it must be to know that you've saved a life, and that eight years later that creature is still alive and well, living free.

June closed with the news that a hen harrier had been found dead in an illegal trap on a grouse moor in Strathbraan, Perthshire. July opened with the disappearance of two satellite-tagged golden eagles on the same morning, again on a grouse moor, again in Strathbraan. Later that month another hen harrier had to be euthanised after its leg was broken in a trap set on a South Lanarkshire grouse moor. Another unsprung trap was found in the harrier's nest.

As each incident hit the press and the list of dead or missing raptors continued to grow, my mind travelled back to January 2017. The reason that those keepers had threatened to picket our wildlife crime talk, and that most within the shooting industry oppose licensing, was now clear to me. The truth is that certain estates are out of control. Their whole business model relies upon total predator elimination. Without that they will never have the grouse numbers they want. Even if licensing does come, are these people likely to change their ways? I have my doubts. And if greater sanctions don't stop the illegal killing – which they won't, unless we find some way of monitoring the hills and properly enforcing the law – the pressure to ban driven grouse shooting will grow. I suspect that the shooting organisations know this. I suspect those keepers do too. I suspect that is why they oppose licensing, why they threatened to picket our talk. Unless the criminals within the industry change their ways, driven grouse shooting will one day be banned in Britain because a ban will be the only sanction remaining, the only one that works. I know that, and those people know it too.

The first weeks of July passed quickly. Mid-month we returned to the osprey nest. As usual, Keith would climb. Dave Anderson had come to fit the rings.

The female rose, spinning in tight circles over the nest, fluting at us as we crossed the field. She'd maintain her vigil all the while that we remained there.

Turning from her to my dad, Keith grinned mischievously. 'What was it you said, Niall? You'll be dead before ospreys nest here...'

Where others might have looked embarrassed, Dad merely laughed.

At the base of the tree Keith's smile slipped. 'Wasn't as dense as this when I put the platform up,' he said, gazing it up and down.

Big, needle-covered branches now ringed the trunk. His climb began slowly and soon grew slower still. Over and over he was forced to stop, unsheathe his saw and hack his way through. By the time he was past the worst of them, my nerves were shredded. I thought of the climbs he'd made last summer, of the nests we'd been sure contained chicks, of the chilled eggs and abattoir scenes we'd found there instead.

Let there be chicks, I thought. Let them be OK.

Somewhere close to the nest he disappeared between another bundle of branches and was lost from sight. I waited and listened, waited, listened.

'All right, Keith?' I shouted. But the wind bore my words away, out from the tree and into the great beyond.

I waited and listened. Waited. Listened.

Then, distant and almost inaudible, came the sound I'd been waiting for.

'Hey! What you saying, eh?'

Every time he sees young birds Keith greets them in this same way.

'How many?' I shouted. And this time the wind was kind. It left my words alone, allowed them to climb up the tree to him.

'Two!' he called back.

'Sending the bag up!'

Minutes later we drew down the zip of the brown hold-all – nominally the possession of Keith's wife, Caroline, but quietly 'appropriated' each summer because it was the perfect size for these jobs. Two lovely speckled osprey chicks sat before us. Almost fully grown, fully realised, they were so very different from any young bird I'd seen before. There was none of the downy fluff that marked the infants of other species. Their legs were thicker and stronger, their talons longer, better designed for grabbing, and for gripping. I thought of young kites, who stare past you through soft grey-blue eyes as they lie motionless on the ground. Their gaze is passive, dreamlike. The ospreys were mild-mannered, too – attempting neither to fight nor to flee – but their flaming orange eyes watched us all the while. These were real birds, old before their time. Next to them every other nestling I'd ever known was just a baby.

As Dave fitted their leg rings, the mother continued to wheel overhead, calling to her chicks. The male soon appeared, too, returning from the hunt. A lifeless fish drooped from his talons. Here on Argaty, where the red kite soars and the red squirrel runs, the children of another recently returned species would eat well that night. What a thought. We returned them to their nest and left them to their meal.

26

A Sign of Things to Come

It was once extinct in England and Scotland. But now the red kite is back with a vengeance as 10,000 take to our skies in the 'biggest species success story in British conservation history'.

'Red Kites Soar Back from the Brink of Extinction', *Daily Mail*, 19 July 2020

A pack of hysterical swallows swirled in late July storms above the farm buildings. Such restless souls, they sped from one point to another, chattering ceaselessly, never stopping anywhere for long.

The British Dragonfly Society had held its annual Scottish members' meeting here the previous week. They visited the raised-bed pond and made approving noises about its construction and the choice of site. However, the general feeling was that if we wished to draw in dragonflies, tall plants were required. In its current state, there was nothing for the nymphs to climb when they emerged from the water. They would, in effect, be trapped down there. So I'd spent the morning raiding our other ponds, seeking appropriate plants for translocation. Now I was planting them in baskets, weighing them down with stones and sending them out like sinking ships to the pond's floor.

The new vegetation oxygenated the water. Inch by inch, minute by minute it began to clear. Within half an hour I could see everything down there. There was my mother's trowel, dropped in by accident one day (I'd send Ross fishing for that, tell him it was part of his apprenticeship); there were the backswimmers, the water snails, the beetles. And there, way down in the depths, so small they were barely visible, were what looked like tiny fish. A closer inspection revealed large frilly gills and four small legs. Baby newts! A frog hopped up the bank, saw me and dived, terrified, for cover. It hit the water with a horsey clop and made straight for the bottom. How satisfying to be surrounded by so much life. I couldn't believe all these creatures had found the pond so quickly.

The air thrummed with the constant buzz of insects. On the banks around the water we had created a small wildflower garden. Of everything we'd sown, the insects' flower of choice was a row of giant poppies. Rowan had scattered the seeds from a pile of old flowers donated by a neighbour. Something about the act of breaking the heads and spreading the contents like salt from a shaker had appealed to her. The results were magnificent. The first blood-red petals had now emerged, bursting forward from those toffee apple heads. How the bumblebees loved them! I watched one, a carder, as she dived in and out of each plant, big yellow balls of pollen stuck to her legs. Hoverflies and wasps came cruising in her wake. This small patch of wildflowers was doing its job wonderfully well.

Sadly the same could not be said for my wildflower meadow. Although we'd followed all of the expert advice, only some of the seeds had taken this year. The dockens had come back in force and crowded many of the others out. Tufted vetch, a hardy grower with pretty purple flowers, had wriggled through; little white campion flowers were glinting here and there. A few ox-eye daisies had sprouted, although their seeds were not in the mix we had purchased. Other than these, the field's main resident was a tall plant with pointed, ovate leaves and a pinkish red head known as redshank or ladies' thumb. We certainly hadn't sown that either (it is generally considered an unwanted weed), but each time I walked the field the wind combed the plants, sending pink waves rippling far and wide, and I couldn't help thinking that something rather pretty lay before me.

Hard, all the same, not to feel a little embarrassed about the failure of so many wildflowers. My parents, on the other hand, were philosophical.

'Who knows what's in the seed bed, and what will emerge next year,' Mum said.

'And besides, those seed heads will be great food for the wee birds when they drop,' Dad added.

The land will do what it wants, not what you want it to do. I shrugged, sighed and resigned myself to this fact.

Of my four wishes for the year, only two had come to pass. The tree sparrows had found the nest boxes and at this very minute our osprey chicks would be standing at the edge of the nest, beating their wings, preparing to fly. Any time now they'd make that great leap of faith. I'd been checking in on them every day, waiting for the moment. Their progress had been one of the high points of the year. But the wildflowers had failed and, for all the things that *had* come, not one dragonfly had found the pond. Nature will do what it wants, not what you want it to do. I shrugged again, sighed again and swallowed more hard truths. In the future, how much time will I waste trying to bend everything to my will? Very little, I suspect. Far better to create something then stand back, let things alone and simply observe.

Much had happened in recent weeks. We'd hosted more talks, run more events. After a hiatus of nearly ten years I'd returned to the world of journalism, contributing articles to local papers, the *Times* 'Thunderer' column and to the Scotland: The Big Picture e-magazine *Think Like a Mountain*. Writing was more challenging than I remembered. In the past decade it seemed I'd lost the ability to put thoughts into words. What a joy to see your work in print, though. I'd forgotten that feeling. It made me think that I'd surrendered my old career too easily all those years ago.

In other respects I was pushing myself, too. In partnership with the British Dragonfly Society we would soon begin hosting dragonfly tours at a pond near the hide.

'Splendid idea,' said Ruary Mackenzie Dodds when I told him the news.

'Do you think we know enough to do it without you?' I asked. Ever since announcing the tours these doubts had been crippling me.

'Your enthusiasm will carry it,' he said, smiling kindly. 'By the way, I think your new pond is fantastic.'

I love that man.

Gill the otter had moved on recently, displaced, I think, by an older female. For a spell they had co-existed without friction. Gill had come in early, the older girl later. Often she'd had a pair of cubs with her. One night I'd caught Gill and the two youngsters on camera at the same time, each gnawing on a fish, and while I'd felt cheered to see them all together I'd known that sooner or later something would have to give. Otters only tolerate one another for so long. Eventually mothers even push their own cubs out of a territory. When that time came, I knew that Gill would go too. This pond was the old girl's and, as long as she was fit, strong and able to drive off rivals, she would be queen of it. Now something had indeed given and Gill was gone, and although the intention had never been to keep her, my heart felt heavy every time I thought of her. It wasn't that I wanted her back – the wild was the best place for her – but I'd have given a great deal to know that she was OK.

I turned my thoughts to happier topics. One of our first

volunteers, Marian White, had kept scrapbooks of the early years of the kite project. Two days earlier I'd found them in a drawer in Mike's old office. The grainy old photographs showed an Argaty that I'd known in childhood – an Argaty I'd long since forgotten. How things had changed. When we'd opened in 2003, we'd had no visitor centre, the patch of land that would become Mike's bumblebee garden was no more than a grassy slope, and everyone looked so much younger! Towards the middle of one book was an article from September 2004, taken from a local newspaper. 'Getting Into A Flap Over Moles' the title read.

Curious, and more than a little perplexed, I read on.

In late August 2004 a local birdwatcher had found a recently fledged kite in the middle of the road, not far from Argaty. Several of the bird's feathers had worn away at the tips. The man had captured the kite and ambulanced it over to Argaty. At little more than 600 grams (even the smallest male fledglings should weigh in at 850 grams), it was very clearly malnourished, so my father had cut some strips of rabbit and put them into the bird's beak. It managed to swallow them but was looking only a little healthier by the next morning. At this point, Dad had employed some peculiar tactics. When dealing with weak lambs, farmers often use the glucose solution Dextrose. Feeling that there was nothing to lose, Dad had coated more rabbit strips in the solution and fed them to the kite. In an instant it had raised its head and begun inspecting its strange new surroundings. Later that day Duncan Orr-Ewing had come to take the bird in. He'd remembered this kite and immediately understood why its feathers were so badly damaged. When they'd climbed the tree to wing-tag the chick, he and his Raptor Group colleagues had found seventeen dead moles decomposing in the nest beside it. Although these had been removed, a further six were discovered in a follow-up visit. How the parents had found so many moles was anyone's guess. Why the chick had rejected their offering was also unknown. Can moles *really* be so much less appetising a prospect than voles, frogs or carrion? Are their hides tougher perhaps, and thus harder to tear into? So many mysteries. One thing was clear: as their bodies had begun to decompose the moles had leaked liquid so acidic it had corroded the poor kite's feathers. When the bird fledged, mangled and underweight, it had never been likely to get far. The Dextrose-marinated rabbit, so said the article, had saved its life.

How could this story have escaped my ears for so long? You'd think my parents and I never spoke to one another! With a smile and a shake of my head, I closed the book.

That smile remained on my face for hours. It's an unusual life we've led.

The last plant due for potting was a lovely, thick marsh marigold. The bright yellow petals, like little beads of sunshine, would add so much colour and vibrancy to the pond's margin. Recently we'd noticed that the water in these shallow parts had been colonised by countless tiny, transparent, syringe-shaped creatures. Our volunteer Sandra, just back from a pond-life workshop, informed me that these were midge larvae. Got water? Have midges. A fundamental Scottish axiom.

Towards the end of the day I returned to admire the new plants and enjoy a few peaceful moments at the pond. I sat watching cabbage white butterflies chase each other through the teasels and poppies, listening to the sound of the bees, thinking that, for all that the wildflower meadow had largely been a disaster, we had, to some small measure, achieved our goal of boosting the insect life of this place. I was sorry about the dragonflies, though. I so wished they'd come.

And then a flash of colour, wine-bottle green and summer sky blue, caught my eye. As little as a year ago I would have been unable to identify an emerald damselfly, now here it was, clear as day, clambering slowly up the marsh marigold stem. Crosshatch cobweb wings glinting, it leapt into the sky and zoomed from plant to plant, marigold to bogbean to towering purple loosestrife. Sometimes its movements were slow and considered. Then some sudden impetus struck and it shot off, traversing the sky at impossible speed. In it came, again and again, searching this place, staking it out, and as I watched it I had a feeling, the same one I'd had eighteen months earlier when we'd found the squirrels and they in turn had found our food. My little winged messenger was here, dancing before me, telling me that something was coming. Something good.

Epilogue

A long fat cloud drifts slowly past, a great old steam train puffing smoke into the grey air. It chugs along on invisible tracks, a soft southwesterly wind guiding it through the darkening sky. I follow its path, watch it climb the steep sides of the hill, slip between storm-broken ranks of conifers on the crest and move onwards, northwards, off to some distant station out there in the night.

In the space vacated by the train fifteen kites are circling. They swirl in a silent commune, one moment flying above me then drifting to the west. Next it's back to the middle ground, then off to the east. Because every movement they make appears practised and controlled, because they fly before the wind, it almost looks as if they are leading it, directing it, taking it by the hand and teaching it how and where to blow. They're above Mike's old house now, moving east towards mine. What do you search for, kites? What do you seek in this thinning light?

It's late September and a change is coming over the land. The days grow shorter, the nights cooler. The squirrels are caching nuts for winter, the wildflowers in the meadow are flopping impotently, our last dragonflies died a few weeks back. Like little birds fledging their nests, the first leaves have abandoned the trees. Summer is on the run, autumn on the march.

In recent days the number of kites roosting here has grown. Once temperatures drop and October sets in our squad will double in size. How reliable these birds are. What comfort they bring. With the world in a mess and the human race pushing it closer and closer towards the cliff edge, you can get lost in the depressing headlines, in the hopelessness of it all. But when everything else is changing the kites keep to their same routine, season upon season, year upon year, utterly impervious to the revolution that gathers steam on their borders.

A realisation struck me recently. Kite protection began in the 1890s. All these years later it's still going strong. This is now the longest continuing conservation initiative in the world. That's quite something. Argaty is just one tiny link in that long chain, but we are a part of some very important history. Everyone who's worked on the project, from the RSPB to the Raptor Study Group and our staff and volunteers, can feel very proud of the work they've done to protect the kites. I feel proud, too.

I think of the fragile set of circumstances that brought kites into our lives, ask myself a series of questions, all beginning *What if . . .* What if Duncan Orr-Ewing had lived somewhere else? What if previous Argaty occupants hadn't unwittingly maintained such kite-friendly habitat? What if the birds had simply chosen to set up camp on another estate? Change but one of these things and our project would never have been, this whole life would never have been. The biggest *what if* of all is the one I save for last. What if my parents had decided against feeding the kites? But I can hardly bear to contemplate that and push the thought away. You can drive yourself mad thinking of these things, picking apart the tenuous threads that somehow knitted together to make you who you are, your story what it is. What are any of our lives, after all, but a combination of circumstances, some happy, some less so? Best just to accept that. A friend recently told me, 'Your parents made a home for the kites and the kites made a home for you.' As far as I'm concerned that's one of the few circumstances that means anything.

'What's your end goal for the estate?' the same friend asked. 'Five or ten years from now, what does the place look like?'

When I began all of this, such questions were far from my mind. I was unhappy and desperate for a way out. My thinking went little beyond that. In light of this, the progress that the project has made these past two years seems pretty remarkable. How did we get this far without a plan? Now, further down the line, I do reflect on the bigger questions. Where *is* all of this headed?

We are so privileged to own this old estate, to be tasked with looking after it in the short time that we have. In his time here, Mike created ponds, a wildflower garden, mile after mile of hedgerow. He started something. In the coming years I'd like to continue it. I want more woodland, more hedgerows, more ponds and more wildflowers. I want an estate where every habitat is linked, healthy and bursting with life. And do you know something? I think we can achieve that. I think we *will*. Now that I've made that pledge to you, fired it out into the great beyond, I'll be too scared to turn back, too scared not to do it.

Beyond that, only a handful of things matter. We'll continue to protect our squirrels, to promote the dragonflies and, most importantly, act as a voice for red kites. I hope that we can convince more people of these birds' brilliance, make a case for them in rewilding circles, stop people killing them. If we achieved only these things and nothing else, it would be enough. I'd consider mine a life well spent.

It's easy to see what we can do to help wildlife on the estate, but tiny, isolated islands of conservation are only so much use. What the world needs is connected habitat, connected thinking. What can we do to influence the situation beyond Argaty? There are a million things, of course, but the main one, I think, is to make our project work. At present, too many people think that the only way to make money from a farm is by traditional means. It doesn't have to be that way. When we first decided to follow this new path, my father gave me a valuable piece of advice: 'If people like you, they'll support you.' I cling to that, treasure it and hope that he's right. If we can make a home for more wildlife and attract more people to come and see it, perhaps we can make a model for others to follow. In turn, perhaps we can change things beyond our home.

Resolved as I am on other fronts, I still haven't entirely made peace with meat production. Will that ever change? I don't know, and maybe that's alright. Maybe we shouldn't ever become too comfortable with what we do, with the sacrifices we demand of the natural world so that we can produce food from it. As I write our cattle are browsing on the hill, trampling down rushes, eating the dead heads of orchids and wildflowers, spreading their seeds through their manure. Meadow pipits and skylarks are digging in those mighty cow pats, feasting on beetles. Next spring another crop of flowers will emerge, cuckoos will return and lay their eggs in meadow pipit nests, a whole cycle of life will start again. Our sheep will give birth too, their placenta will feed a new generation of kites. These stories are so often overlooked, drowned out by the negative press that hangs over farming nowadays, but they matter. They prove that human survival needn't always be about taking from our world. Sometimes, if we do it right, we can give something back. I'm trying not to overlook them. My stance for Argaty is: farm well, make a place for nature, do the best you can. That is all anyone can ask of us. As for doubting my motivations for ever embarking upon this journey, I'm testing a new tactic: worrying less about *why* I'm doing what I'm doing, and trying to remember that what matters most is not *why* you try to do the right thing but simply *that* you try to do it.

*

Above my head the kites are spinning, running laps in the sky, moving east towards the Drum. As so often happens, Charlie's journals enter my mind. I read them again recently. Midway through I found the following paragraph, dated 1930, which I'd previously overlooked:

> I have found out from *Woods, Forests and Estates of Perthshire* by Hunter 1833 that the Drum was planted out about 200 years ago. I wonder when the planting we are doing now will be cut down.

A few pages later came this, from 1932:

> The ride along the ridge running approximately north and south has been planted out as a special feature. Limes are planted out every six yards, the centre of the ride being in direct line with the Wallace Monument. It is hoped that someday this may be of interest.

These two small extracts say so much. Through them you see a man fully aware that he was only a small part of this place, a chapter in its story, a man who planted trees knowing he'd be dead before they reached maturity, who made decisions for a future he wouldn't see.

Though Charlie didn't divulge the location of the limes, I knew the trees would be found in the Drum. No other wood is even remotely close to the Monument. When I visited, I found them within minutes. There they stood, a whole row of them stretching north to south, patiently waiting for the day that they might prove of interest. And on the hills beyond, the Wallace Monument loomed tall. The trees stood in a perfect line before it, just as Charlie said. I'm grateful for everything that my great-grandfather and subsequent generations did for Argaty, for the home they made for me. I hope I can care for it as well as they did.

The night is growing colder. Any residual warmth is leaving

the air. Looking north, I gaze past the fields, the hide, the woods where the first kites nested, towards Gill's pond. Where are you now, Gill? How I'd love to know. I turn to the south. Somewhere out there in the immensity of this night are four Argaty ospreys en route to Africa. Perhaps they've arrived already. The chicks left the nest in early August. Their mother promptly absconded, beginning her long journey south, but the father stuck by them, taught them the ways of the wild. Now they are all gone and we have begun counting down the days to next spring, hoping the parents return safely. It's sure to be a long, nervous wait.

Above my head fifteen kites are spiralling, ambling really, in no particular rush to go anywhere, and faced with such a sight one can only be happy. They're turning in silent circles, making kaleidoscopic patterns on the clouds, and the beauty and serenity of this moment hits me right in the heart. I want to capture it, bottle it, experience it over and over again.

Scenes like this are worth fighting for, worth saving. That these birds continue to thrive in this hard, hard world ought to give us hope. History has taught us how vulnerable they are, how much they rely on our goodwill. We wiped them out once before and, were we minded to, could easily do so again. Instead their numbers are growing. That means something. To me kites are a reflection of the changes we are making as a society. They symbolise our improvement.

The light is dropping now, the sky changing from one thing to another, and the birds are going, heading for the trees, dropping in a chain into the canopy. I keep up my watch until the last of them disappears, and then I begin my descent down the hill, to Sarah and Rowan and home.

You have read many quotes in this book. Some were charming, some harrowing, some a combination of the two. I sought long and hard for the right one to finish my story with. Then, turning on the radio one day, I heard a warm, familiar voice. David Attenborough was narrating an episode of Radio 4's 'Tweet of the Day', and as I listened I felt tears straining my eyes. I had found what I'd been looking for.

I leave you with it.

> On cold winter evenings dark wraiths circle the woods and squeal in the gathering gloom. Red kites are arriving at a roost. The presence of red kites in so many places in the UK would have been unthinkable a few decades ago. If you wanted to see this graceful, fork-tailed bird you'd have to travel to the remote valleys of mid-Wales where the last few pairs hung on. After centuries of persecution, red kites were almost wiped out by gamekeepers, but in 1989 a project to restore the red kite back into the wild began with a release of birds in the Chiltern Hills and in Scotland. Since then there have been more introductions and kite numbers have soared so that now these birds are foraging even around the outer suburbs of London. The return of the red kite to our lowland landscapes is a conservation triumph, and when you see them floating over their woodland roosts on a winter's evening, remember that we almost lost them forever.

> – 'Tweet of the Day', BBC Radio 4, 26 December 2013

Postscript

After a successful first summer of tours, Argaty was awarded the title of Dragonfly Hotspot by the British Dragonfly Society in August 2019.

In November that year the Scottish Government's Independent Grouse Moor Review Group published their report, recommending additional regulation for practices including muirburn, mountain hare culls and the use of medicated grit on grouse. On grouse moor licensing they were split. A unanimous decision was reached that a licensing system should be introduced if, after five years, 'there was no marked improvement in the ecological sustainability of grouse moor management, as evidenced by populations of breeding Golden Eagles, Hen Harriers and Peregrines on or within the vicinity of grouse moors being in favourable condition'.

Between February and March 2020 we ran a birdbox-making project with the children of nearby Doune and Deanston primary schools. Each of the 220 pupils built a nest box and took it home to install in their garden. More than thirty friends helped with the construction of the boxes. We were truly humbled by their support.

Later in March the coronavirus (COVID-19) pandemic brought the UK to a standstill. The respiratory disease swept through the country, hospitals were overrun, tens of thousands of people died. To slow the spread of the virus, schools and non-essential shops were closed and people were made to stay at home. In line with government guidelines we shut our doors to the public but continued to work to improve habitats on the estate.

We established four new wildlife ponds, taking our total to sixteen. Backswimmers, lesser water boatmen and diving beetles quickly took up residence in each; bats hunted over them at night.

By April a huge crop of red campion had emerged in our wildflower meadow. Come summer, white campion, meadow buttercup, knapweed and viper's bugloss had also flowered. Our friend Sandra Stewart, from the advisory group Farming and Conservation, approached us with a plan to sow a further half-acre of species-rich grassland in the field behind the red kite hide. Buoyed by the success of the first patch we accepted. The new plot came good immediately. Cornflowers, yarrow, bird's-foot-trefoil and ox-eye daisies shot up, and little rainbows of colour rippled through the field. By July the dockens had crowded out many of the flowers in the original meadow again. For the umpteenth time I shrugged and said something about nature doing what it wanted, not what we wanted it to. But I was encouraged to note that there *was* life in there. Spiders spun webs between the weeds, songbirds were foraging there too, and on late summer mornings when the sun hung low over the land the dew-drizzled webs glistened like endless strings of fairy lights, charms of goldfinch flickered and flashed between drooping seed heads, and I had to conclude that this was, in an unexpected way, working.

By late spring the COVID infection rate had levelled out and schools and businesses were allowed to reopen on a limited basis. Although fewer tourists came to Scotland throughout the summer, our squirrel hide, nature walks and kite tours somehow sold out every day. Lockdown had had some unexpected impacts. Life had slowed down. Freed from the commitments of school, commutes and the mad rush of modern life many people found the time to stop and really notice the natural world around them. It was a silver lining to a very hard year.

Gill the otter was not seen here again.

The ospreys returned for a second successive year, rearing three chicks.

Over the course of the summer five different dragonfly species bred in our raised bed pond.

In August, a week shy of my 37th birthday, I shot my first grey squirrel. This brought me no pleasure. Nor, I admit, did it bring me sorrow. Such things are beyond me now. In truth, I think they have been for a long time. But I knew my reasoning, knew what I sought to protect. My mind was clear.

In November the Scottish Government issued its response to the Grouse Moor Review Group Report. A year had passed since the report was published. In that time one hen harrier had been shot, another six had disappeared, a white-tailed eagle had died after ingesting a banned poison and a golden eagle had vanished, all on Scottish grouse moors. A shot buzzard had also been found dead on a sporting estate and a man had been charged with the poisoning of over twenty birds of prey in Dumfries and Galloway. The government response read:

> Since 2007, the Scottish Government has undertaken
> a range of measures to tackle wildlife crime . . . The
> fact that raptor persecution continues in spite of all
> the measures we have already taken suggests that, while
> regulation from within the grouse shooting industry can
> be an important factor, self-regulation alone will not be
> enough to end the illegal killing of raptors and further
> government intervention is now required.

A grouse moor licensing system would be introduced within the next parliamentary term. Mountain hare culls and muirburn would also be subject to increased legislation.

The red squirrels continue to delight us with their antics. They are now to be found all over the estate, regularly stealing peanuts (and destroying bird feeders) in each of our neighbours' gardens. Between 2019 and 2020 we built two more squirrel hides.

Scotland's kite population continues to grow. There are now almost 400 breeding pairs across the country. What progress they've made.

Thanks

So many people have helped us on this journey. To everyone who's supported us along the way, thank you. We're so grateful.

In particular I owe thanks to:

Alan Jones, Fiona Brims and Marion Moore for the wonderful days with the squirrels. Those are some of the happiest memories of my life.

Andrea Hudspeth, Kari de Koenigswarter and Ruary Mackenzie Dodds for sharing your love of dragonflies with us.

Andrew O'Donnell for many fun days out filming wildlife around the farm.

Bob McGowan and the team at National Museums Scotland's Department of Natural Sciences. Thank you for steering my research in the right direction.

Charlie Bowser, for the things you did for Argaty, and for taking the time to record your story. How little we'd know of history without people like you.

Cian O'Driscoll, for the many nights spent discussing good music, mediocre rugby, terrible football and, occasionally, our respective books. Thank you for being so thoughtful and generous with your time and ideas. I miss you, my friend.

Colin Seddon, for trusting us with Gill. She was one of the great privileges of our lives.

David Anderson, Ian Thomson, Peter Cairns and Stewart Miller. Your thoughts helped shape these pages.

Deborah Warner, Andrew Simmons and all at Birlinn, for all of your help and support.

Duncan Orr-Ewing, Lorcan O'Toole, Kevin Duffy and Simon Zisman, the RSPB's Central Scotland red kite team, for everything you have done, and continue to do, for these wonderful birds.

Gordon Buchanan, both for the lovely foreword to this book and all the amazing animal advocacy you do. Thank you.

Judith Midgley, Nicola Horsburgh, Ràna Webster, Richard Cuthill, Ross Graham, Ryan Pownall, Sandra McDerment and Talia Vilalta Capdevila for all of your sterling pond-building work. What a wonderful thing we created.

Keith Burgoyne, ringing buddy, friend and climber extraordinaire. Thank you for the many great days out in the wild. Here's to many more.

Lorenzo Ranalli for reading through various early parts of this book.

Mike McDonnell. Whatever happens the rest of our lives, wherever we go, whatever friends we meet, nobody will be able to replace you and all the things we went through. How I wish we'd taken these next steps together.

Our wonderful volunteers – Clare, Gavin, Gwyneth, John, Laurent, Marian White, Michael, Steve, Vicki and Victoria – for being the heart and soul of our project. We couldn't do this without you.

Philip Price. Always positive, always good to bounce ideas off. Thanks for the many discussions on kites, nature and life.

Polly Pullar, without whom this book would never have happened. You set this up, read the book in its various drafts and your feedback was ever-helpful. Thank you for believing in me and in this project.

Roy Dennis, for bringing the first kites back to Scotland and England. What kind of world would we live in without people like you?

Lastly, to my parents, Lynn and Niall, who worked so hard to make a home for the birds and a future for me. Words can't express my pride or gratitude.

And to Sarah and Rowan. Strong and beautiful, you enrich the world around you. I love you both very much.

Select Reading

Books

Booth, E.T. *Rough notes on the Birds observed during twenty-five years shooting and collecting in the British Islands* (Vol. I), R.H. Porter, London (1887)

Carrick, J.D., Motherwell, W., Henderson, A. & Robertson, D. *The Laird of Logan: Anecdotes and Tales Illustrative of the Wit and Humour of Scotland*, David Robertson & Co., Edinburgh (1889)

Carter, I. *The Red Kite*, Arlequin Press, Chelmsford (2001)

Colquhoun, J. *The Moor and the Loch*, William Blackwood & Sons, Edinburgh (1840)

Conway, J. *Forays Among Salmon and Deer*, Simpkin, Marshall, Hamilton, Kent & Co., Glasgow (1899)

Forbes, R.A. *Gaelic Names of Beasts (Mammalia), Birds, Fishes, Insects, Reptiles, etc.*, Oliver & Boyd, Edinburgh (1905)

Gow, D. *Bringing Back the Beaver*, Chelsea Green, London (2020)

Harvie-Brown, J.A. *A Fauna of the Tay Basin & Strathmore*, David Douglas, Edinburgh (1906)

Harvie-Brown, J.A. & Buckley, T.E. *A Vertebrate Fauna of Sutherland, Caithness and West Cromarty*, David Douglas, Edinburgh (1887)

Harvie-Brown, J.A. & Buckley, T.E. *A Vertebrate Fauna of Argyll and the Inner Hebrides*, David Douglas, Edinburgh (1892)

Harvie-Brown, J.A. & Buckley T.E. *A Vertebrate Fauna of the Moray Basin* (Vols I & II) David Douglas, Edinburgh (1895)

Harvie-Brown, J.A. & Macpherson, Rev. H.A. *A Vertebrate Fauna of the North West Highlands and Skye*, David Douglas, Edinburgh (1904)

Hunter, T. *Woods, Forests, and Estates of Perthshire, with Sketches of the Principal Families in the Country*, Henderson, Robertson & Hunter, Perth (1883)

Jamieson, J. *Jamieson's Dictionary of the Scottish Language*, W.P. Nimmo, Edinburgh (1867)

Lovegrove, R. *The Kite's Tale: The Story of the Red Kite in Wales*, RSPB, Bedfordshire (1990)

Lovegrove, R. *Silent Fields: The Long Decline of a Nation's Wildlife*, Oxford University Press, Oxford (2007)

Mactaggart, J. *The Scottish Gallovidian Encyclopedia* (second edition), Thomas J. Morrison & Co., Glasgow (1876)

Montagu, G. *Ornithological Dictionary of British Birds*, Orr and Smith, London (1831)

Pullar, P. & McIntyre, N. *The Red Squirrel, A Future in the Forest*, SCOTLAND: The Big Picture (2017)

Ritchie, J. *The Influence of Man on Animal Life in Scotland*, Cambridge University Press (1920)

Sim, G. *A Vertebrate Fauna of the Dee*, D. Wyllie & Son, Aberdeen (1903)

St John, C. *Short Sketches of the Wild Sports and Natural History of the Highlands*, John Murray, London (1849)

St John, C. *A Scottish Naturalist: The Sketches and Notes of Charles St John 1809–1856*, Andre Deutsch, London (reprinted: 1982)

Reports

Analyses of the Fate of Satellite Tracked Golden Eagles in Scotland, Scottish Natural Heritage, Commissioned Report No. 982. Edinburgh (2017)

The State of Nature Reports (2013 & 2016)